Elementary Structural Analysis and Design of Buildings

A Guide for Practicing Engineers and Students

Elementary Structural Analysis and Design of Buildings

A Guide for Practicing Engineers and Students

Dominick R. Pilla

CRC Press
Taylor & Francis Group
Boca Raton London New York

CRC Press is an imprint of the
Taylor & Francis Group, an **informa** business

CRC Press
Taylor & Francis Group
6000 Broken Sound Parkway NW, Suite 300
Boca Raton, FL 33487-2742

First issued in paperback 2018

© 2017 by Taylor & Francis Group, LLC
CRC Press is an imprint of Taylor & Francis Group, an Informa business

No claim to original U.S. Government works

ISBN-13: 978-1-4987-7588-5 (hbk)
ISBN-13: 978-0-367-02804-6 (pbk)

Visit the Taylor & Francis Web site at
http://www.taylorandfrancis.com

and the CRC Press Web site at
http://www.crcpress.com

Contents

10 Foundations and retaining structures   195

11 Structural review of construction   241

About the author

Dominick R. Pilla is an engineer and architect, working in the industry and as an associate professor at the School of Architecture, The City College of New York. Professor Pilla completed his undergraduate study at Rensselaer Polytechnic Institute, Troy, New York and earned his MS in civil engineering at New Jersey Institute of Technology and continues to conduct independent research at The City College of New York.

Professor Pilla has served as principal-in-charge of all of Dominick R. Pilla Associates, Professional Corporation's projects since the firm's inception in 1999. As a result of his training and experience as both an engineer and an architect, he is aware of the influence of materials that affect analysis and design of structures.

Drawn from Professor Pilla's teaching experience at The City College of New York and his work as a design engineer and an architect, he has developed *Elementary Structural Analysis and Design of Buildings*, a comprehensive guide and desk reference for practicing structural and civil engineers and for engineering students.

Introduction

This book is an introduction to the process of building engineering as performed by professional structural engineers. To gain the required knowledge and to properly engineer buildings, it is common to be formally educated in engineering, and then to take part in an apprenticeship as a junior engineer where the professional practice is learned during work experience. The junior engineer is taught to navigate the facets of building design by applying those principles taught at school with professional practice standards. This book allows the reader to link the theory with practice and illustrates typical applications used in everyday practice. The process presented in this book covers industry standard applications and interpretations of required building codes as well as the use of building code-adopted design references for the analysis and design of buildings. While the material presented in this book is at an elementary level, its example-based presentation is at a professional level and can be thought of as a simple road map for similar contextual situations.

Building design is often thought to consist of those systems that are gravity supporting, such as columns and beams, and lateral resisting, such as shear walls and frames. It is the lateral forces, specifically the seismic requirements due to the anticipated seismic forces, which limit the structural system selection and dictate the required detailing for a building. For this reason, the subject matter discussed in this book is largely based on the lateral system analysis and design of buildings.

The process of professionally engineering a building must address the following topics:

- Minimum design loads for buildings
- Wind and seismic forces applied to buildings
- Lateral force distribution
- Discussion of simplified analysis methods
- Design and detailing of structures
- Steel, concrete, wood, and masonry lateral systems
- Foundations and retaining structures

A brief discussion of building code requirements pertaining to structural inspections is also covered in this book to give the reader an appreciation of the required quality control measures to ensure a properly built structural system.

This book is not intended to be all inclusive in regard to the principles and practice of engineering design of buildings. It is meant to provide a linear progression of concepts and how they fit within the design process. The reader is assumed to have a basic working knowledge of design and is encouraged to use the codes and design standards referenced in this book in conjunction with completing the problems presented. The objective is to gain the confidence to apply these principles to the other structural systems not discussed.

How to use this book

This *Elementary Structural Analysis and Design of Buildings* guide is intended for professionals (engineers and architects), for students of architecture and engineering, and for those interested in gaining a thorough understanding of the process of engineering design of buildings. The reader should be able to use this book as a primer to the sequence of planning as it relates to the engineering design of buildings. It can be used as a standalone reference or as a text for instruction on the engineering process. The subject matter is presented to the reader in a systematic sequence, which allows the reader to understand the basic topics and build upon them with each chapter.

This text is current with the applicable material design references and building codes at the time it was published. Tables, figures, and excerpts of text are summarized from the design references and building codes cited in the "Codes and References" chapter of this book, so the reader is able to seamlessly follow the examples and progression of subjects without having to stop and reference the industry codes and design guides. However, the reader is encouraged to review the applicable codes and references to obtain a thorough understanding of the subject matter presented in this book and how it appears in the references.

This book is divided into 11 chapters that progress from the description and application of loads experienced by buildings to the analysis of structures and to the engineering design of buildings and their components, consistent with the industry practice.

Chapter 1

Minimum design loads for buildings

1.1 LOADS

The structural system of a building is designed to sustain or resist anticipated loads or forces the building may experience during its life in order to provide a reliably safe building structure. Engineers and architects use building codes, which have been developed based on statistical data, to aid designers with the basis of calculating the required loads. However, it cannot be overstated that the building designer must recognize the potential loads and apply them correctly for analysis.

Typical loads imposed on a building are vertical loads, such as dead and live loads, and lateral loads, such as wind and seismic and lateral earth pressures. Building structures experience many other additional loads such as loads due to thermal and hydrostatic forces. We will review these loads and others in more detail in Chapters 2, 3, 4, and 10. However, to understand loads, we want to discuss some of the basics about the loading of buildings.

A building's vertical loading is based on its intended use, the number of occupants and the type of construction, and which are the building's dead and live loads, respectively. Dead loads depend on the materials used to construct the building, and live loads are based on the anticipated occupants using the building. Loads are often applied in combination based on their likelihood of occurring simultaneously. Determining the appropriate load to use for structural analysis and design requires knowledge of the long and short duration loads. For example, a warehouse has a much higher floor load than an office or a residential building because of the weight of the contents of the storage in the warehouse, contributing to its dead load, as compared to that of an office or a residential building, which generally has more occupants and therefore a higher live load. In this case, the storage is long term and the occupants are transient. Building codes take this into account and consider the appropriate statistical loading to be used in structural calculations. The type of materials and construction will also determine loading by altering the building's weight or mass. A two-story steel and concrete building, for instance, is likely to be considerably heavier than a wood-framed building of the same size. However, an early circa 1900s masonry building with flat-arch floor construction is heavier yet. The materials selected are consequential in determining the dead load of a building.

A building's location will dramatically affect its loading and consequentially its structural system also. A building located in Buffalo, New York, for instance, will experience much higher snow loads than a building in New York City due to the potential accumulations of lake-effect snowfall in the Great Lakes region of the United States. Similarly, a building located on the West Coast of the United States as compared to a building on the East Coast will experience much higher seismic loading due to a much more active ground motion on the West Coast. Or a building located near the coastline will experience higher

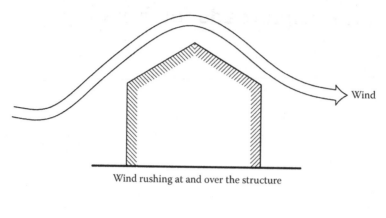

Wind rushing at and over the structure

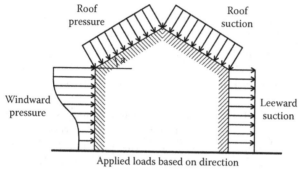

Applied loads based on direction

Figure 1.1 Wind pressure on building surfaces.

wind forces than a structure inland that is protected by surrounding buildings, trees, and other topographical characteristics.

Wind rushing over a building with a gable roof, as shown in Figure 1.1, experiences wind forces on all surfaces of the structure. Consequently, the building's primary structural system or main wind force-resisting system is designed to resist these forces. In addition, various loads are applied in combination based on their likelihood of occurring simultaneously.

The loads considered for the design of buildings are called minimum design loads and are in accordance with local and national building codes. The International Building Code (IBC) references the "Minimum Design Loads for Buildings and Other Structures" published by the American Society of Civil Engineers (ASCE) and is the standard for determining applied loads on a building to be used for structural analysis and design.

1.2 DEAD LOADS

Dead load is the self-weight of the building that is composed of all the construction materials that form the building: the roof, floors, walls, foundations, stairs, mechanical components, plumbing and electrical fixtures, built-in cabinetry and partitions, finishes, cladding, and all permanent equipment. Or simply put, if you were to imagine turning a building upside down and shaking it, everything that did not fall out would be considered the dead load. The typical composite steel floor section, shown in Figure 1.2, illustrates building components, which contribute to self-weight.

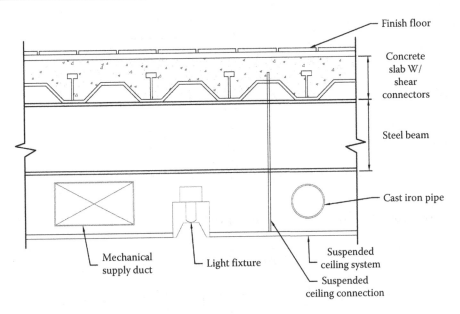

Figure 1.2 **Composite steel construction floor section.**

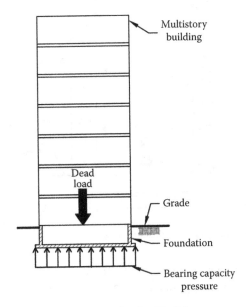

Figure 1.3 **Resulting bearing pressure due to gravity loads of building.**

Obviously the dead load of a building is extremely important. A building's structural system must be able to support its self-weight (dead load) as well as all other possible loads the building may experience. The foundations, which support the weight of the building, must transfer its load to the supporting soils or rock, which the building bears on (see Figure 1.3).

The dead load contributes to the stability of a buildings' structure. Heavier buildings or structures are able to resist lateral loads by pure mass. The retaining wall in Figure 1.4 experiences a lateral load from the soil pressure it supports. The weight of the wall resists the overturning force by counteracting with its weight.

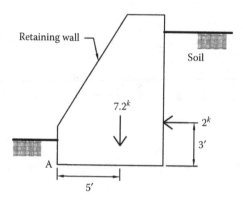

Figure 1.4 Weight of wall and lateral force on retaining wall.

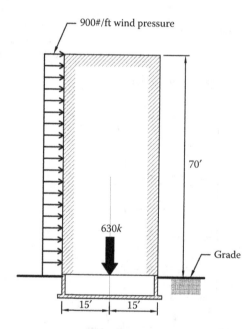

Figure 1.5 Lateral wind loading and gravity load of building.

$$\text{Factor of safety} = \frac{\text{Righting moment}}{\text{Overturning moment}} = \frac{(7.2k)(5')}{(2k)(3')} = 6$$

Similarly, the lateral wind load pressure on the building in Figure 1.5 is resisted by the weight of the building. The factor of safety is calculated by dividing the righting moment by the overturning moment as shown.

$$\text{Factor of safety} = \frac{\text{Righting moment}}{\text{Overturning moment}} = \frac{(630k)(15')}{(.900k)(70')(35')} = 4.3$$

which means the building is stable by a factor of safety of 4.3.

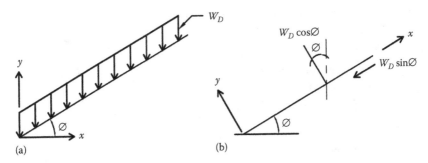

Figure 1.6 (a) Global and (b) local coordinate systems.

The dead load of a sloping member is shown in Figure 1.6. The weight of the member acts in the global coordinate system. To analyze the member, the dead load force must be normalized to its local coordinate systems as shown.

1.3 LIVE LOADS

The live loads used in the design of a building are the maximum loads imposed by the occupants using the building. That is to say, for example, the anticipated live load imposed on a structure, for residential use, will differ to that compared to an office building or school, and the live load is less because of fewer occupants for a residential use. Based on theoretical and statistical data, a compiled list of design loads has been assembled. It is well understood and accepted among practitioners that the tabulated design loads listed are conservative; actual values of live loads, when surveyed, are usually less. Listed in Table 1.1 are the minimum uniformly distributed live loads based on occupancy or use. For a complete listing of both uniformly distributed and concentrated live loads see ASCE 7, Table 4.1.

The components of buildings, such as the roof, walls and floors are to be designed to sustain uniformly distributed live loads or concentrated live loads placed such that they produce the maximum load effect in the member.

1.3.1 Reduction in uniform live loads

According to ASCE 7, the design uniform live load can be reduced except for those members supporting roof uniform live loads. Member live load reduction has been an accepted practice since the 1960s. The methodology has evolved, and the permitted reductions are based on the following formula and criteria:

$$L = L_o \left[0.25 + \frac{15}{\sqrt{K_{LL}A_T}} \right] \tag{1.1}$$

where:

L is reduced design live load per ft² of area supported by the member, (lb/ft²)
L_o is unreduced design live load per ft² of area supported by the member, (lb/ft²)
K_{LL} is live load element factor (see Table 1.2)
A_T is tributary area in ft²

A member, having a tributary area (A_T) multiplied by its live load element factor (K_{LL}), resulting in at least 400 ft², is permitted to have its live load reduced according to Equation 1.1.

Table 1.1 Minimum uniformly distributed live loads

Occupancy or use	Uniform psf
Hospitals	
Operating room, laboratories	60
Patient room	40
Corridors above first floor	80
Libraries	
Reading rooms	60
Stack rooms	150
Corridors above first floor	80
Manufacturing	
Light	125
Heavy	250
Office buildings	
Lobbies and first floor corridors	100
Offices	50
Corridors above first floor	80
Residential	
Private rooms and corridors	40
Public rooms and corridors	100
Roofs	
Flat, pitched and curved	20
Roofs used as gardens	100
Schools	
Classrooms	40
Corridors above first floor	80
First-floor corridors	100
Stairs and exit ways	100
Stores, retail	
First floor	100
Upper-floor	75
Wholesale; all floors	125

Table 1.2 Live load element factor, K_{LL}

Element	K_{LL}
Interior columns	4
Exterior columns w/o cantilever slabs	4
Edge columns with cantilever slabs	3
Corner columns with cantilever slabs	2
Edge beams with cantilever slabs	2
Interior beams with cantilever slabs	2
Cantilever beams with cantilever slabs	1
One- and two-way slabs	1

Note: For a complete listing of live load element factors, see ASCE 7-10, Table 4.2.

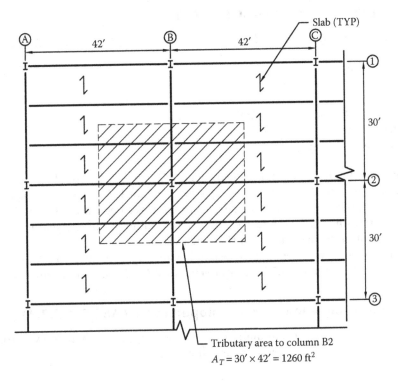

Figure 1.7 Partial floor framing plan.

Example 1.1
The steel and concrete slab partial floor plan, as shown in Figure 1.7, is that of a typical floor in an office building. From Table 1.1, the live load for an office building is 50 psf. To determine if a live load reduction is permitted, the term $K_{LL}A_T$ must be at least 400 ft².

Obtaining the live load element factor, K_{LL}, from Table 1.2 and calculating the tributary area of column B2, the term $K_{LL}A_T = 4$ (1260 ft²) = 5040 is much greater than 400 ft², and therefore the column is permitted to be designed for a reduced live load.

Hence, $L = L_o \left[0.25 + \dfrac{15}{\sqrt{K_{LL}A_T}} \right] = 50 \left[\dfrac{0.25 + 15}{\sqrt{(4 \times 1260)}} \right]$ psf = 35.6 psf

1.4 SNOW LOADS

Building roofs must be structurally designed to sustain loads imparted by snow. The structural engineer must design structural systems of roofs to sustain snow loads for all of the states in United States with the exception of Florida. The entire state of Florida has a mapped ground snow load of zero. The mapped snow loads in ASCE 7 are based on the historical data associated with recoding ground snow depths. The mapped ground snow loads, p_g, for the contiguous 48 states of the United States is found in ASCE 7-10 (Figure 7.1) and is used to calculate roof snow loads.

1.4.1 Flat roof snow loads (ASCE 7, 7.3)

The flat roof snow load, p_f, is calculated using the following formula, in (lb/ft²):

$$p_f = 0.7C_eC_tI_sp_g \tag{1.2}$$

where:
C_e is exposure factor, given in Table 1.3
C_t is thermal factor, given in Table 1.4
I_s is importance factor given in Table 1.5

The ground snow load, p_g, is obtained from Figure 7.1 in ASCE 7-10. The exposure factor, C_e, in Table 1.3, is correlated to terrain categories B, C, or D for the site, which correspond to exposure categories B, C, and D and surface roughness categories B, C, and D. For design purposes, the terrain category and roof exposure condition chosen should represent the anticipated condition during the life of the structure.

Surface roughness categories and exposure categories are defined in Chapter 26 "Wind Loads," Sections 26.7.2 and 26.7.3, respectively, in ASCE 7-10, and are summarized here.

1.4.2 Minimum snow load for low sloped roofs (ASCE 7, 7.3.4)

The code, ASCE 7, requires a minimum roof snow load, p_m, and shall apply to roofs having a slope of less than 15°. The criteria are as follows:

Where the ground snow load, p_g, is 20 lb/ft² or less: $p_m = I_sp_g$.
Where the ground snow load, p_g, is greater than 20 lb/ft²: $p_m = 20 (I_s)$.

Surface Roughness Categories
Surface Roughness B: Urban and suburban areas, wooded areas, or other terrains with numerous closely spaced obstructions having the size of single-family dwellings or larger.
Surface Roughness C: Open terrain with scattered obstructions having heights less than 30 ft. This category includes flat open country and grasslands.
Surface Roughness D: Flat, unobstructed areas and water surfaces. This category includes smooth mud flats, salt flats and unbroken ice.
Exposure Categories
Exposure B: For buildings with a mean roof height of less than or equal to 30 ft, exposure B shall apply where the ground surface roughness, as defined by surface roughness B, prevails in the upwind direction for a distance greater than 1500 ft. For buildings with a mean roof height greater than 30 ft, exposure B shall apply where the ground surface roughness, as defined by surface roughness B, prevails in the upwind direction for a distance greater than 2600 ft or 20 times the height of the building, whichever is greater.
Exposure C: Exposure C shall apply for all cases where exposure B or D does not apply.
Exposure D: Exposure D shall apply where the ground surface roughness, as defined by surface roughness D, prevails in the upwind direction for a distance greater than 5000 ft or 20 times the height of the building, whichever is greater. Exposure D shall also apply where the ground surface roughness immediately upwind of the site is B or C, and the site is within a distance of 600 ft or 20 times the height of the building, whichever is greater, from an exposure D condition as defined in the previous sentence.

Table 1.3 Exposure factor, C_e

| | Exposure of roof | | |
Terrain category	Fully exposed	Partially exposed	Sheltered
B	0.9	1.0	1.2
C	0.9	1.0	1.1
D	0.8	0.9	1.0
Above tree line in mountainous areas	0.7	0.8	N/A

Use Exposure Categories for Terrain Categories shown in Table 1.3.

Example 1.2

A three-story office building located on Main Street in Nyack, NY, in close proximity to the Hudson River (Figure 1.8) has an upwind direction from the river. In order to calculate the flat roof snow load for the building, we need to determine the variables in Equation 1.2.

Solution

Step 1: Obtain the ground snow load (p_g) from Figure 7.1 in ASCE 7-10. The ground snow load for Nyack, NY is 30 psf, ($p_g = 30$ psf).

Step 2: Determine the roof exposure. The exposure factor, C_e, is based on the wind exposure of the building and the surface roughness. The building, as shown in Figure 1.8, is 40 ft tall, and has a surface roughness in compliance with that of surface roughness B. The surface roughness prevails for a distance of approximately 2200 ft, which is less than the 2600 ft required to satisfy the condition for exposure category B. Additionally, exposure category D is not satisfied and consequently exposure category C prevails. Hence, use terrain category C in Table 1.3, and for a fully exposed roof, $C_e = .9$.

Step 3: Determine the thermal factor, C_t, which is based on the thermal condition of the building described in Table 1.4, hence $C_t = 1.0$.

Step 4: Determine the importance factor, I_s, which is based on the risk category assignment of the building in Table 1.5-1, in ASCE 7-10. Risk categories I, II, III, and IV are based on the potential loss of life during a catastrophic failure. According to Table 1.5-1, an office building has a risk category of II. From Table 1.5, a risk category II has an importance factor, $I_s = 1.0$.

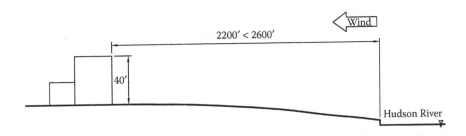

Figure 1.8 Wind direction and distance from body of water.

Table 1.4 Thermal factor, C_t

Thermal condition	C_t
All structures except as indicated below	1.0
Structures keep just above freezing and others with cold, ventilated roofs in which the thermal resistance (R-value) between the ventilated space and the heated space exceeds 25°F × h × ft/Btu	1.1
Unheated and open air structures	1.1
Structures intentionally kept below freezing	1.2
Continuously heated greenhouses with a roof having a thermal resistance (R-value)	1.3
less than 2.0°F × h × ft²/Bt	0.85

Table 1.5 Important factors by risk category of buildings for snow, ice, and earthquake loads

Risk category from Table 1.5-1	Snow importance factor, I_s	Ice importance factor–thickness, I_i	Ice importance factor–wind, I_w	Seismic importance factor, I_e
I	0.80	0.80	1.00	1.00
II	1.00	1.00	1.00	1.00
III	1.10	1.25	1.00	1.25
IV	1.20	1.25	1.00	1.50

Step 5: Finally, the flat roof snow load for the building in Figure 1.8 is found from Equation 1.2.

$$P_f = 0.7 C_e C_t I_s p_g$$

$$P_f = 0.7(.9)(1.0)(1.0)(30 \text{ psf}) = 18.9 \text{ lb/ft}^2$$

1.4.3 Snow drifts on lower roofs (ASCE 7, 7.7)

Step roofs will form snowdrifts depending upon the roof configuration and the direction of the wind in relation to the roofs. Stepped roofs can accumulate drifting on either the leeward or windward side of an upper roof. That is, snow blown from an upper roof onto a lower roof (the lower roof is on the leeward side of the upper roof) will accumulate in a drift on the lower roof. Also snow on a lower roof, which is blown against the wall of a building forming an upper roof (the lower roof is on the windward side of the upper roof), will form a drift (see Figure 1.9).

The height of the balanced snow load, h_b, is calculated by dividing the snow load by the snow density.

$$h_b = \frac{p_s}{\gamma} \tag{1.3}$$

where:
  p_s is the weight of the snow
  γ is the snow density

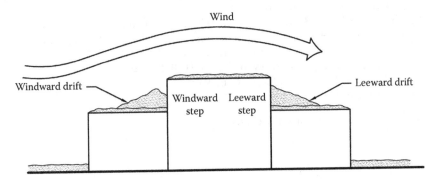

Figure 1.9 Wind-driven snow drifts on lower roofs: Leeward and windward drifts.

The snow density is found as follows:

$$\gamma = .13(p_g) + 14 \tag{1.4}$$

Example 1.3
Calculate the snowdrift on the lower roof for the stepped roof residential building shown in Figures 1.8 and 1.10. The ground snow load (p_g) is 30 psf. The building is located in an urban, mixed use vicinity, consisting of residential and commercial buildings.

Solution
Step 1: Calculate the high roof flat roof snow load (see Example 1.1).
Exposure category C

$C_e = .9$ (Fully Exposed)

$C_t = 1.0$

Occupancy category II

$I_s = 1.0$

$p_f = 0.7C_eC_tI_sp_g$

$p_f = 0.7(.9)(1.0)(1.0)(30 \text{ psf}) = 18.9 \text{ lb/ft}^2$

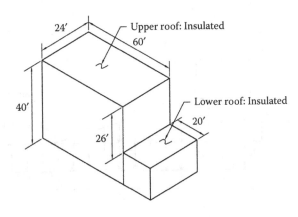

Figure 1.10 Steeped roof building.

Check the minimum roof snow load.

Because $p_g = 30$ lb/ft², is greater than 20 lb/ft², p_m, is equal to the importance factor, I_s, times 20. Hence, the minimum roof snow load, $p_m = (1.0)\ 20 = 20$ lb/ft², is greater than the calculated flat roof snow load of 18.9 lb/ft².

Hence, the minimum roof snow load, $p_m = (1.0)\ 20 = 20$ lb/ft², controls.

Step 2: Calculate the low roof flat roof snow load.

Exposure category C

$$C_e = 1.0 \qquad \text{(partially exposed)}$$

$$C_t = 1.0$$

Occupancy category II

$$I_s = 1.0$$

$$p_f = 0.7 C_e C_t I_s p_g$$

$$p_f = 0.7(1.0)(1.0)(1.0)(30\ \text{psf}) = 21.0\ \frac{\text{lb}}{\text{ft}^2}$$

Step 3: Calculate the snow density and height of the balanced snow.

The weight of the snow, p_s, is then equal to the flat roof snow load, p_f, and the snow density, γ, is equal to, $0.13(30) + 14 = 17.9$ lb/ft³, using Equation 1.4.

$$\text{Hence, } h_b = \frac{p_s}{\gamma} = \frac{21\ \text{lb/ft}^2}{17.9\ \text{lb/ft}^3} = 1.17\ \text{ft}$$

Step 4: Determine if a drift load is required.

Calculate, h_c, as shown in Figure 1.11, by subtracting the height of the balanced snow load, h_b, from the stepped roof height.

Hence, $h_c = 26\ \text{ft} - 1.17\ \text{ft} = 24.83\ \text{ft}$

If h_c/h_b is less than 0.2, drift loads are not required to be applied.

$h_c/h_b = 24.83\ \text{ft}/1.17\ \text{ft} = 22.22$, which is much greater than 0.2, therefore drift loads must be applied to the lower roof.

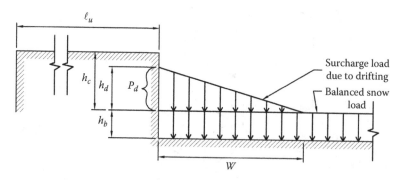

Figure 1.11 Configuration of snow drifts on lower roofs.

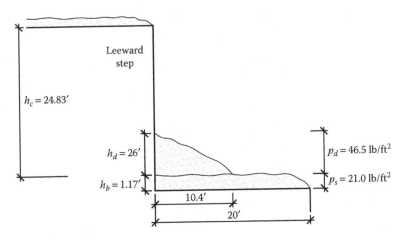

Figure 1.12 Leeward drift dimensions and loading on the lower roof.

Step 5: Determine the height, h_d, of the leeward drift.

Using ASCE 7, Figure 7-9, "graph for determining drift height," enter the graph with a ground snow load, p_g, equal to 30 lb/ft² and l_u, the length of the building (upper roof) parallel to the drift, equal to 60 ft. The corresponding drift height, h_d, is equal to 2.6 ft.

Step 6: Determine the height, h_d, of the windward drift.

Using ASCE 7, Figure 7-9, "graph for determining drift height," enter the graph with a ground snow load, p_g, equal to 30 lb/ft² and l_u, the length of the building parallel (lower roof) to the drift, equal to 20 ft. The corresponding drift height, h_d, is equal to 1.7 ft.

Hence, the leeward drift height, h_d, equal to 2.6 ft, controls.

Step 7: Calculate the width, w, of the drift and loading, p_d.

If the height of the drift, h_d, is less than, h_c, which is the height of the adjacent wall to the drift, then the width, w, is equal to $4(h_d)$.

Hence, h_d, is much less than, h_c, (2.6 ft < 24.83 ft) and, $w = 4$ (2.6 ft) = 10.4 ft. See ASCE 7, Section 7.7-1, for complete description of determining drift width on lower a roof.

And finally, to calculate the snow drift loading, p_d, multiply the height of the drift, h_d, by the snow density, γ.

Hence, p_d, is equal to (2.6 ft)(17.9 lb/ft³) = 46.5 lb/ft².

See Figure 1.12 for dimensional configuration and loading of the drift on the lower roof of the building.

1.5 THERMAL LOADING

All buildings and their components are subjected to temperature changes and consequentially thermal loadings. The response of a structure to thermal loadings can be crucial to the wellbeing of the structure and should be considered in design and detailing of the structure. Thermal loadings can put stress onto the building and can create cracks in concrete and masonry and can fail structural steel if not considered properly.

When an object is heated or cooled, its length will change by an amount proportional to the original length and the change in temperature. All materials have a coefficient of linear thermal expansion. Some more common construction materials and their associated coefficient of linear thermal expansion, α, are listed here.

Material	$\alpha(10^{-6}\,in/in°F)$
Aluminum	12.3
Concrete	8.0
Glass (plate)	5.0
Granite	4.0
Masonry	2.6–5.0
Mortar	4.1–7.5
Plastics	22.0–67.0
Slate	5.8
Steel	7

Linear thermal expansion of an object can be expressed as

$$\Delta l = \alpha \Delta t \ (L_0)$$

(1.5)

where:

Δl is change in length of object (in.)
α is coefficient of linear thermal expansion (10^{-6} in/in °F)
Δt is change in time $= (t_1 - t_0)$

where:

t_0 is initial temperature (°F)
t_1 is final temperature (°F)
L_0 is initial length of the object (in.)

Example 1.4

The steel canopy shown in Figure 1.13, is anchored to the building on the left and is exposed to a temperature range of −10°F to 110°F during the year. Calculate the anticipated change in length at the end of the canopy and determine if the gap provided between the canopy and the building to the right is adequate.

Solution

The length change of the canopy is calculated as follows:

$$\Delta l = \alpha \Delta t \ (L_0)$$

(1.5)

$$= 6.7 \times 10^{-6} \left[110°F - (-10°F) \right] \left(100\ ft \times \frac{12\ in}{ft} \right) = 0.965\ in$$

which is less than the 2 in. gap provided, so the gap or space between the canopy and the building is adequate.

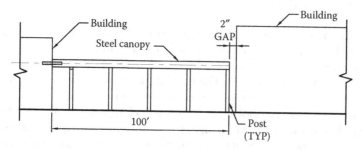

Figure 1.13 Elevation of building with a steel canopy attached.

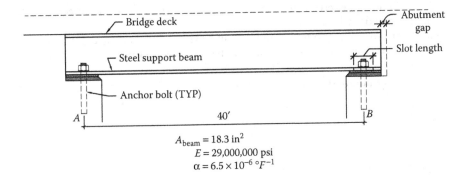

Figure 1.14 Elevation of steel and concrete bridge.

Example 1.5

A steel and concrete bridge is supported on masonry abutments as shown in Figure 1.14. The steel support beams bear on steel bearing plates secured to the masonry abutments by anchor bolts.

The bridge support beams shown are exposed to a temperature range of −20°F to 100°F ($\Delta t = 120°F$).

To calculate the anticipated length change of the beam and determine the required slot length and gap length of abutment B is calculated as follows:

Solution

$$\Delta l = \alpha \Delta t \ (L_0) \tag{1.5}$$

$$= 6.7 \times 10^{-6}(120°F)\left(40 \ \text{ft} \times \frac{12 \ \text{in}}{\text{ft}}\right) = 0.386 \ \text{in}$$

Therefore, specify a gap greater than 0.386 in. and a slot length of 7/8 in., which is approximately twice the length change to accommodate thermal growth in either direction.

Not providing a slot in the beam, discussed in Example 1.5, would result in a force applied to the bolt, due to the thermal elongation of the beam. The bolt, which is embedded in the masonry of the abutment, would experience a shear force through the shank. If we want to quantify the magnitude of the force applied to the bolt, we need to consider stress and strain relationships as well as Hooke's Law.

Stress is defined by force over an area,

$$\sigma = \frac{P}{A} \tag{1.6}$$

and Strain is defined by the displacement of a member divided by the original length of that member,

$$\varepsilon = \frac{\delta}{L} \tag{1.7}$$

Then using Hooke's law, in the form of stress equal to the modulus of elasticity multiplied by the strain,

$$\sigma = E\varepsilon \tag{1.8}$$

we can assemble and express, in either terms of deformation or load, an equation for deformation of axially loaded members.

Hence, rearrange Equations 1.7 and 1.8 and express in terms of deformation and strain, respectively, as $\delta = \varepsilon L$, and $\varepsilon = \sigma/E$, can be combined to form

$$\delta = \varepsilon L = \frac{\sigma L}{E} \text{ and substituting } \sigma = \frac{P}{A}, \text{ yields}$$

$$\delta = \frac{PL}{AE} \tag{1.9}$$

Example 1.6

Consider the support beam shown in the steel and concrete bridge, supported on masonry abutments, in Figure 1.14, and calculate the shear force across the anchor bolt shank due to the axial elongation of the beam under the thermal loading described Example 1.5.

Solution

Rearrange Equation 1.9 in terms of load as $P = \delta AE/L$ and using the axial deformation calculated in Example 1.5, the resulting force is

$$P = \frac{(0.37 \text{ in})(18.3 \text{ sq in})(29000 \text{ ksi})}{\left[40 \text{ ft} \times (12 \text{ in/ft})\right]} = 409 \text{ kips}$$

This force is obviously very large and completely unnecessary to try to resist. Consequently, thermal forces must be identified and accounted for in the design of almost every type of structure.

1.6 FORCES AND LOADS DUE TO SOIL PRESSURES

Soil, which is against a foundation or retaining wall or any vertical surface, imposes a lateral soil pressure onto that surface. Lateral earth pressure is exerted on structures situated in soil, such as foundation walls and earth retaining structures. The depth of the structure, the slope of the soil at grade, the type of soil, the placement of the soil against the structure, presences of ground water and the type of structure are all factors in determining design pressures.

1.6.1 Active and passive lateral pressure

The reinforced concrete cantilever retaining wall shown in Figure 1.15 retains 17.5 ft of soil, the difference between the top of the grade and the bottom of the footing. This is the height of the soil, which will laterally load the wall. The vertical reinforcement in the wall is designed to resist the bending of the wall from the lateral soil pressure. The front of the footing has 3 ft of soil against it. This soil will contribute to the stability of the retaining wall by providing resistive loading.

We can refer to retaining structures having an active and passive side of the wall. Soil placed against the retaining wall, on the high side, is said to be active because as the wall moves or bends due to the lateral loading pressure of the soil. The wall will be engaged by the soil moving toward the wall, which is considered to be the active lateral soil pressure. Similarly, the soil in front of the wall provides resistance or pressure against the structure, thereby preventing the structure from

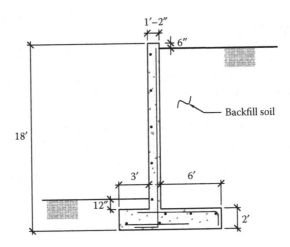

Figure 1.15 Section of reinforced concrete cantilever retaining wall.

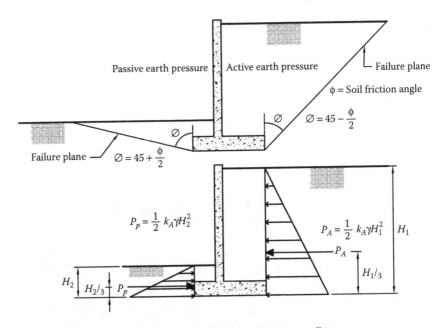

Figure 1.16 Active and passive earth pressure on cantilever retaining wall.

sliding forward, this is considered to be the passive lateral soil pressure. The active and passive soil wedges move along failure planes, shown in the top illustration of Figure 1.16.

Another way of thinking of active and passive pressures is to imagine the active wedge of soil, being placed or backfilled against the retaining wall, loading the wall laterally as the backfill is placed up to the top of the wall, the wall begins to bend and allow the soil to slide along its failure plane. The active wedge of soil then imposes pressure against the stem and the entire structure. It not only bends the stem, but pushes the wall and promotes sliding of the structure also, which in turn engages the soil in front of the retaining structure and causes the passive soil wedge to slide along its failure plane.

The failure plane shown in Figure 1.16 is a function of the soil friction angle, which is a soil property of the specific type of soil used.

The corresponding passive and active soil pressures are triangle distributions as shown in the bottom illustration of Figure 1.16.

Triangular pressure distributions, shown, are simply calculated as half the base times the height of the soil. The base of the distribution is equal to height of the wall times the unit weight of the soil times a lateral soil pressure coefficient (KYH).

The active and passive lateral soil coefficient, K_a and K_p, are calculated using the Rankine earth pressure theory. Hence, the area of the triangular distribution equals the lateral earth pressure on the vertical surface of the wall.

The lateral earth pressure, acting at 1/3 the height of the wall is calculated as

$$P = \frac{1}{2} K Y Y H \tag{1.10}$$

Example 1.7

For the retaining wall shown in Figure 1.15, consider the earth-pressure coefficients, and the soil weight given and calculate the active and passive pressures on the wall.

Active pressure coefficient. $K_a = .333$
Passive pressure coefficient, $K_p = 3.33$
Soil weight, $Y = 120$ pcf

Solution

These forces are then used to determine structural stability. It is a common practice to ignore the contribution of passive soil pressure to thereby provide a conservatively designed system in the event the passive soil is compromised during the life of the retaining structure (Figure 1.17).

Figure 1.17 Retaining structure installed against in-place embankment shoring. Shoring braces are removed after wall has cured and reached full strength.

1.6.2 Static lateral soil pressure

Soil which has been deposited naturally and consolidated (compacted under its own weight) normally overtime is said to be "at-rest" or "static" and is considered not to be moving along a failure plane as seen with active and passive soil states. Additionally, at-rest lateral soil coefficients, K_o, are between that of active and passive coefficients. At-rest lateral coefficients range from 0.40 for dense sand to 0.60 for loose sand.

Soil placed against a foundation with a basement slab and first floor framing as shown in Figure 1.18, is commonly designed as a static lateral soil pressure condition. This is because the basement slab and the first floor provides bracing at the top and the bottom of the wall. The foundation wall, is essentially laterally supported by the slab and the floor and the wall bends between the two supports the way a floor would bend except rotated on its side (the wall does not move laterally).

Example 1.8
Consider the earth-pressure coefficients, and the soil weight given for the foundation walls shown in Figure 1.19, and calculate the static pressure on the wall.

Soil type: dense sand

Soil weight, $\Upsilon_{s\,dry} = 120$ pcf

Soil weight, $\Upsilon_{s\,submerged} = 145$ pcf

Static pressure coefficient. $K_o = 0.40$

$\Upsilon_{water} = 62.4$ pcf

Solution
$F_1 = \frac{1}{2}$ (240 psf) (5 ft)^2 = 3000 #
$F_2 = $ (240 psf) (9 ft) = 2160 #
$F_3 = \frac{1}{2}$ (522 psf) (9 ft)^2 = 21,141 #
$F_4 = \frac{1}{2}$ (562 psf) (9 ft)^2 = 22,761 #

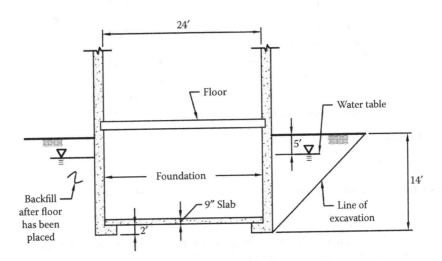

Figure 1.18 Section of a building foundation with respect to grade and water table.

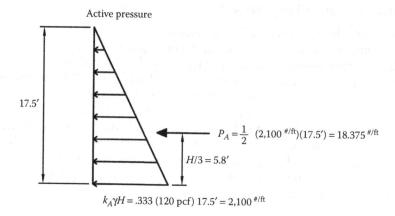

$$P_A = \frac{1}{2}\,(2,100\;^{\#/ft})(17.5') = 18.375\;^{\#/ft}$$

$$H/3 = 5.8'$$

$$k_A\gamma H = .333\;(120\text{ pcf})\;17.5' = 2,100\;^{\#/ft}$$

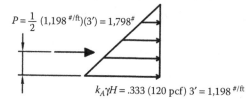

$$P = \frac{1}{2}\,(1,198\;^{\#/ft})(3') = 1,798^{\#}$$

$$k_A\gamma H = .333\;(120\text{ pcf})\;3' = 1,198\;^{\#/ft}$$

Figure 1.19 Force diagrams of active and passive soil pressures.

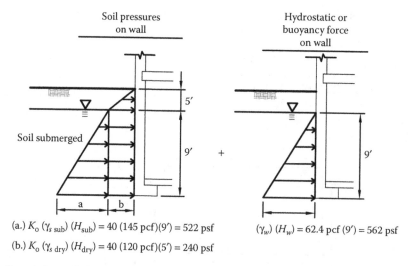

(a.) $K_o\,(\gamma_{s\;sub})\,(H_{sub}) = 40\,(145\text{ pcf})(9') = 522$ psf

(b.) $K_o\,(\gamma_{s\;dry})\,(H_{dry}) = 40\,(120\text{ pcf})(5') = 240$ psf

$(\gamma_w)\,(H_w) = 62.4\text{ pcf }(9') = 562$ psf

Figure 1.20 Foundation soil pressures.

Therefore the total force on the foundation wall is the sum of the forces:

$$F_1 + F_2 + F_3 + F_4 = 49,062\;\#$$

The foundation wall can then be analyzed for the four forces applied to the wall in their respective locations (Figures 1.20 and 1.21).

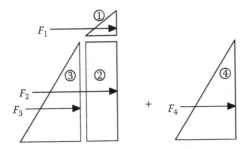

Figure 1.21 Free body diagrams of soil pressures.

1.6.3 Hydrostatic pressure

Not only are pressures applied to the sides of foundations due to the presence of water but the foundations and slabs experience buoyancy forces as well. The slab of the building shown in Figure 1.22 has a buoyancy or hydrostatic force applied to the underside of the basement slab. This force exerts an upward pressure on the slab due to the foundation's slab being submerged 7 ft.

The designer must consider the upward force on the slab to properly provide design specifications for a slab which will sustain this loading. In addition, waterproofing measures must be developed to keep water from entering the structure. A slab designed for a hydrostatic pressure considers the weight of slab to counteract the upward force.

Example 1.9
The net uplift on the slab due to buoyancy is

$$= \Upsilon_{water}(H_{water}) - \text{weight of slab}$$

$$= 62.4 \text{ pcf } (7 \text{ ft}) - 9/12 (150 \text{ pcf}) = 324.3 \text{ psf}$$

So, the slab must span the width of the building and sustain the net upward force on the slab. If this pressure is not identified and properly designed, the slab will fail.

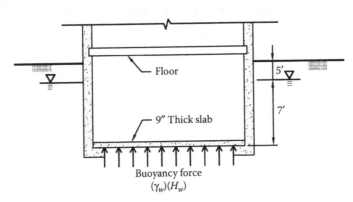

Figure 1.22 Section of a building foundation with upward buoyancy force applied to underside of slab

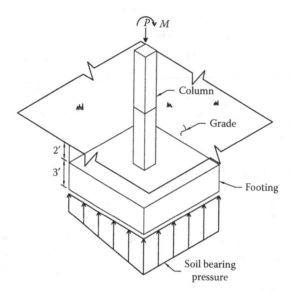

Figure 1.23 Spread footing with a column load and overturning moment.

1.6.4 Bearing pressure

The adequacy of the soils a foundation bears on, will determine the size and the type of the footing. If soils are present which will support spread footings then the footing's size will be dependent upon the loads it can sustain. Typically, a spread footing is sized for bearing capacity and any overturning moments. The footing shown in Figure 1.23 has an axial load as well as an overturning moment applied to the column/pier. The footing transfers the loads to the soil and creates an upward reaction on the underside of the footing as shown. The pressure distribution is in response to the applied loads that is, it is equal and opposite.

Note that the pressure distribution is greater at one end of the footing than the other. This is because the overturning moment is additive to the column's axial load, thereby yielding a higher bearing force at one end. The capacity of the soil must be able to sustain the higher value or the footing will have to be increased to spread the load over a larger area to decrease the bearing force and bring it into an acceptable range.

This concept of sizing the footing based on the applied loads and bearing capacity of the soil will be discussed in sufficient detail in Chapter 10, Foundations and Retaining Structures.

Chapter 2

Wind and seismic forces applied to buildings

2.1 LATERAL LOADS

In addition to gravity loads (dead and live loads), buildings are designed to resist lateral loads, predominantly in the form of wind, seismic (earthquake), and lateral soil pressures. Buildings are designed to respond to lateral loads that are exerted on them, by generating resistance. The internal member forces, which are generated by the applied loads, must resolve through an explicit load path to the building's foundation and ultimately to the earth. A building's lateral structural system is specifically designed to meet the requirements of minimum design loads as prescribed by the governing building code.

The purpose of building codes is to provide standardized guidelines, which are aimed to protect public health, provide safety and general welfare as they relate to the construction, and occupancy of buildings and structures. For this reason, the International Code Council, which produces the International Building Code, adopts the American Society of Civil Engineers (ASCE) 7 standard: Minimum Design Loads for Buildings and Other Structures. This design reference standard provides the minimum permitted design loads and load combinations for both strength design (load resistance factor design) and allowable stress design (ASD).

2.2 WIND LOADS

Designing structural systems to sustain wind loads is a relatively new concept in regard to structural design of buildings. As the building sciences and technologies advanced, essentially beginning with the industrial revolution, the ability to design lighter and lighter structures became possible. This is directly related to an advantageous decrease in building costs due to the consequential conservation of building materials and a decrease in the required labor to construct buildings. As a result of designing building structures with less mass, buildings become susceptible to wind pressures and they then must be designed to sustain anticipated wind pressures to prevent failure.

Wind loads on buildings are transmitted through the building's cladding and then to the components, which transfer the load to the building's main wind force-resisting system (MWFRS). The outer surface or skin of the building, such as curtain wall systems, roof coverings, and exterior windows and doors, is typically considered to be cladding. The cladding pressure, as determined using ASCE 7, is different: higher than the pressure generated for the MWFRS. Components are considered to be fasteners, purlins, girts, studs, roof rafters, and decking and roof trusses. The design wind pressure is the same for both components and cladding. The MWFRS is a frame or shear wall system, which may consist of an assemblage of structural elements that work together to resist lateral wind loads acting on the structure and direct them to the foundations and then to the ground. Components can also be part of the MWFRS when they act as shear walls or roof diaphragms.

The format for wind loading, as presented in the ASCE 7 standard, is covered in the following chapters:

- Chapter 26: Wind Loads, General Requirements
- Chapter 27: MWFRS (Directional Procedure)
- Chapter 28: MWFRS (Envelope Procedure)
- Chapter 29: Components and Cladding
- Chapter 30: Wind Loads on Other Structures and Appurtenances
- Chapter 31: Wind Tunnel Procedures

2.2.1 Directional procedure

The directional procedure, as described in Chapter 27, of the ASCE 7-10 standard, for developing the MWFRS wind loading is presented in this chapter. This method is based on past wind tunnel testing and of prototypical building models. The external pressure coefficients have been derived from the testing performed and correspond to the direction of the wind applied to the building.

The following topics are the required components to generate the minimum required loading according to the ASCE standard.

2.2.2 Surface roughness categories (ASCE, Section 26.7.2)

Surface roughness B: Urban and suburban areas, wooded areas, or other terrain with numerous closely spaced obstructions having the size of single-family dwellings or larger.

Surface roughness C: Open terrain with scattered obstructions generally less than 30 ft. This category includes flat open country and grasslands.

Surface roughness D: Flat and unobstructed areas and water surfaces. This category includes smooth mud flats, salt flats, and unbroken ice.

2.2.3 Exposure categories (ASCE, Section 26.7.3)

The exposure categories are controlled by the surface roughness categories as follows:

Exposure category B: Where surface roughness B prevails upwind for ≥ 2600 ft or 20 times the building height (h).
  Exception: ≥ 1500 ft, if $h \leq 30$ ft.

Exposure category D: Where surface roughness D prevails upwind for ≥ 5000 ft or 20 times the building height (h).
  Also, where the surface roughness immediately upwind of the site is B or C and the site is within a distance of 600 ft or 20 times the building height (h).

Category A: None.

Category C: Applies when exposures B and D do not apply.

2.2.4 Velocity pressure (ASCE, Section 27.3.2)

The velocity pressure, q_z (psf), is a dynamic pressure and is a function of the 3-s gust basic wind speed V (mph), the height of the building, the terrain, and the directional coefficients:

$$q_z = .00256 \, K_z K_{zt} K_d V^2 \qquad\qquad \text{(2.1) (ASCE Equation 27.3-1)}$$

K_z or K_h = velocity pressure exposure coefficient (Table 2.1 [ASCE T 27.3.1])
K_{zt} = topographical factor (ASCE Figure 26.8.1)
K_d = wind directional factor (Table 2.2 [ASCE T 26.6-1])
V = basic wind speed (ASCE Figure 26.5.1A, B, or C)

2.2.5 Internal pressure

The internal pressure of a building is based on the openings and obstructions inside the building. Buildings are classified as enclosed, partially enclosed, or open. A building is *enclosed* when wind is not permitted to enter the building, that is, the building envelope

Table 2.1 Velocity pressure exposure coefficients, K_h and K_z

Height above ground level, z		Exposure (Note 1)		
Ft	(m)	B	C	D
0–15	(0–4.6)	0.57	0.85	1.03
20	(6.1)	0.62	0.90	1.08
25	(7.6)	0.66	0.94	1.12
30	(9.1)	0.70	0.98	1.16
40	(12.2)	0.76	1.04	1.22
50	(15.2)	0.81	1.09	1.27
60	(18)	0.85	1.13	1.31
70	(21.3)	0.89	1.17	1.34
80	(24.4)	0.93	1.21	1.38
90	(27.4)	0.96	1.24	1.40
100	(30.5)	0.99	1.26	1.43
120	(36.6)	1.04	1.31	1.48
140	(42.7)	1.09	1.36	1.52
160	(48.8)	1.13	1.39	1.55
180	(54.9)	1.17	1.43	1.58
200	(61.0)	1.20	1.46	1.61
250	(76.2)	1.28	1.53	1.68
300	(91.4)	1.35	1.59	1.73
350	(106.7)	1.41	1.64	1.78
400	(121.9)	1.47	1.69	1.82
450	(137.2)	1.52	1.73	1.86
550	(152.4)	1.56	1.77	1.89

Table 2.2 Wind directionality factor, K_d

Structure type	Directionality factor K_d
Buildings	
Main wind force-resisting system	0.85
Components and cladding	0.85
Arched roofs	0.85
Chimneys, tanks, and similar structures	
Square	0.90
Hexagonal	0.95
Round	0.95
Solid signs	0.85
Open signs and lattice framework	0.85
Trussed towers	
Triangular, square, rectangular	0.85
All other cross sections	0.95

including the windows and doors are designed to withstand cladding pressures and consequently has a low internal pressure. A *partially enclosed* building is one where the openings allow the wind to enter and then encounter obstructions, which prevent the wind from leaving and thereby create high internal pressures. This is the case when the windows blow in or fail due to high wind pressure; the wind entering the building creates an internal pressure on the interior surfaces causing failures from the inside out of the building. An *open* building allows the wind to enter and leave relatively easily without encountering many obstructions and has no internal pressure; see Figure 2.1.

Internal pressure, p_i (psf), is determined as follows:

$$p_i = \pm \, q_i(GC_{pi}) \qquad\qquad\qquad (2.2)\ (\text{ASCE Equation 27.4-1})$$

$q_i = q_h$ = velocity pressure is calculated at $z = h$, which is the mean roof height (feet).
GC_{pi} = internal pressure coefficient (Table 2.3 [ASCE T 26.11-1]).

2.2.6 Gust-effect factor (ASCE, Section 26.9)

The gust-effect factor is based on whether the building is rigid or flexible. In order to determine whether a building is flexible or rigid, the fundamental natural frequency must be established using the structural properties and deformational characteristics of the resisting elements in a properly substantiated analysis. A building that has a fundamental natural frequency less than 1 Hz is considered to be flexible.

For a rigid building, the gust-effect factor is conservatively permitted to be taken as 0.85. A low-rise building is an enclosed or partially enclosed building that has a mean roof height (h) less than or equal to 60 ft. In addition, the mean roof height (h) does not exceed the least horizontal dimension. Taller buildings with more specific data can be determined rigid or flexible, using the equations of ASCE 7, Section 26.9.

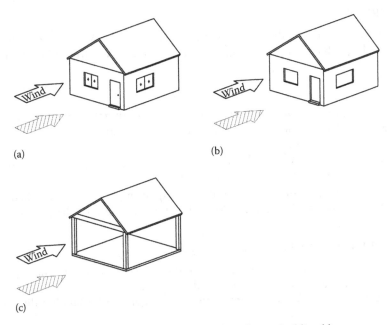

(a)   (b)

(c)

Figure 2.1 Illustration of enclosed (a), partially enclosed (b), and open building (c).

Table 2.3 Internal pressure coefficient, GC_{pi}

Enclosure classification	GC_{pi}
Open buildings	0.00
Partially enclosed buildings	+0.55
	−0.55
Enclosed buildings	+0.18
	−0.18

2.2.7 External pressure coefficient (ASCE, Figure 27.4-1)

The external pressure coefficients have been developed specifically for the windward, leeward, sidewalls, and roof with respect to the direction of wind under consideration.

2.2.8 Design pressure

Then, the design wind pressure for the MWFRS for enclosed or partially enclosed rigid buildings, for all heights, is determined by the following equation:

$$p = qGC_p - q_i(GC_{pi}) \qquad\qquad (2.3)\ \text{(ASCE Equation 27.4-1)}$$

$q = q_z$ for positive pressures (windward walls at height z above the ground)
$q = q_h$ for negative pressures (leeward walls, side walls, and roofs at height h)

2.2.9 Parapets

The design wind pressure for the effects of the parapets on the MWFRS of rigid or flexible buildings with flat, gable or hip roofs shall be determined by the following equation:

$$P_p = q_h(GC_{pa}) \qquad\qquad (2.4)\ (ASCE\ Equation\ 27.4\text{-}4)$$

P_p = is the net pressure on the parapet due to the combination of the net pressures on the surfaces from the front and the back of the parapet.

The velocity pressure, q_h, is evaluated at the top of the parapet and GC_{pa} is 1.5 for windward parapet and −1.0 for leeward parapet (Figure 2.2).

The pressures on the parapet as determined by Equation 2.4 are used in lieu of the pressures calculated by Equation 2.3 for the MWFRS, for the height of the parapet.

Example 2.1: Determine the MWFRS velocity pressure

Determine the MWFRS velocity pressure (q) at the mean roof height ($h = 60$ ft), and the design wind wall pressures. For an office building located in Montauk, NY, with an exposure category C site, neglect topographic effects; see Figure 2.3 for building dimensions.

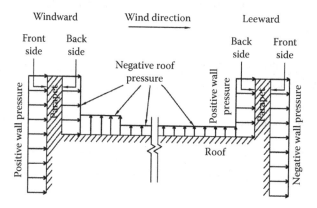

Figure 2.2 Wind pressures on windward and leeward parapets and roof.

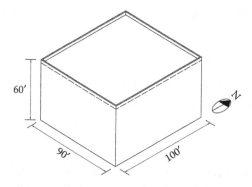

Figure 2.3 Building dimensions of office building located in Montauk, NY.

Step 1: Determine the risk category from ASCE T 1.5-1 *Risk Category of Buildings and Other Structures for Flood, Wind, Snow, Earthquake and Ice Loads.*

Because an office building is not designated as an essential facility or a facility which if fails will cause a substantial impact on civilian life, it is not risk category III or IV. Also an office building is not designated as a low risk to human life in the event of a failure, so it would not be risk category I. According to Table 1.5-1, a building that is not classified as risk categories I, III, or IV, is assigned to risk category II.

Step 2: Determine the basic wind speed, V, for the applicable risk category, using ASCE Figure 26.5-1A, B, or C.

The basic wind speed, V, using Figure 26.5-1A Basic Wind Speeds for Risk Category II Buildings and Other Structures, for Montauk, NY is 140 mph.

Therefore, $V = 140$ mph.

Step 3: Determine the load parameters,

Wind directional factor: $K_d = 0.85$ (Table 2.2)

Exposure category: C (given in problem)

Topographical factor: $K_{zt} = 1.0$ (given in problem)

Gust-effect factor: $G = 0.85$ (ASCE Section 26.9)

Enclosure classification: enclosed building

Internal pressure coefficient: $GC_{pi} = \pm 0.18$ (Table 2.3)

Step 4: Determine the velocity pressure exposure coefficient,

Height of the building: $h = 60$ ft

Exposure coefficient: $K_z = K_{60} = 1.13$ (Table 2.1)

Step 5: Determine the velocity pressure,

$$\text{Velocity pressure}: \quad q_z = .00256\, K_z K_{zt} K_d V^2 \tag{2.1}$$

$$q_z = .00256\, K_z (1.0)(0.85)(140)^2 = 42.65\, K_z \text{ psf}$$

$$q_z = q_{60} = 42.65\, K_z = 42.65\,(1.13) = 48.19 \text{ psf}$$

z (ft)	K_z	q_z (psf)
60	1.13	48.19
50	1.09	46.49
40	1.04	44.36
30	0.98	41.80
20	0.90	38.39
10	0.85	36.25

Step 6: Determine the external pressure coefficients (ASCE 7 Figure 27.4-1) (Figure 2.4),

N–S motion:

Horizontal dimension of the building, normal to wind direction: $B = 90$ ft

Horizontal dimension of the building, parallel to wind direction: $L = 100$ ft

$$\frac{L}{B} = 1.11, \text{ then,}$$

Wall pressure coefficient: Windward wall: $C_p = 0.8$

Leeward wall: $C_p = -0.3$

Side Wall: $C_p = -0.7$

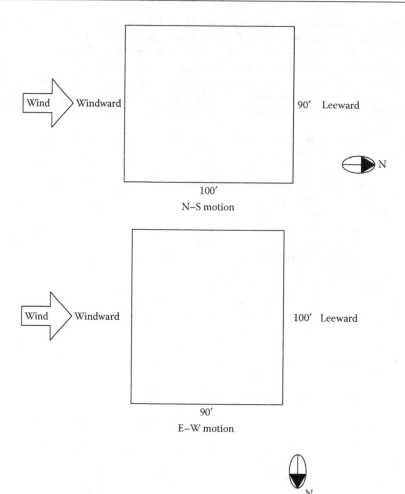

Figure 2.4 **Plan view of building with wind applied to windward side for N–S motion and E–W motion.**

E–W motion:
   Horizontal dimension of the building, normal to wind direction: $B = 100$ ft
   Horizontal dimension of the building, parallel to wind direction: $L = 90$ ft

$$\frac{L}{B} = 0.9, \text{ then,}$$

Wall pressure coefficient: Windward wall: $C_p = 0.8$
                            Leeward wall: $C_p = -0.5$
                            Side wall: $C_p = -0.7$
Roof:
Flat roof: $\Theta = 0$ (degrees)
$h/L = 0.6$ or 0.67 is greater than 0.5, and less than 1.0, therefore interpolation for
   C_p is required (Table 2.4).
Step 6: Calculate the wind pressure, p, on each wall surface (windward and leeward):

$$p = qGC_p - q_i(GC_{pi}) \tag{2.3}$$

Table 2.4 Wall pressure coefficient, C_p

Roof Cp:		N–S motion	
Distance from windward edge	$h/L \leq 0.5$	$h/L = 0.6$	$h/L = 1.0$
0 to h/2	-0.9	-0.98	-1.3
h/2 to h	-0.9	-0.86	-0.7
h to 2h	-0.9	-0.86	-0.7
		E–W motion	
Distance from windward edge		$h/L = 0.67$	
0 to h/2		-1.03	
h/2 to h		-0.83	
h to 2h		-0.83	

The external pressure is the first part of the equation, and is as follows:

$$p_e = qGC_p$$

Then, the pressures for the north–south direction can be calculated as follows:

$$p_{e\,windward} = 48.19(0.85)(0.8) = 32.77 \text{ psf}$$

$$p_{e\,leeward} = 48.19(0.85)(-0.3) = -12.29 \text{ psf}$$

Similarly, the pressures for the east–west direction:

$$p_{e\,windward} = 48.19(0.85)(0.8) = 32.77 \text{ psf}$$

$$p_{e\,leeward} = 48.19(0.85)(-0.5) = -20.48 \text{ psf}$$

The internal pressure is calculated with the last part of Equation 2.3 as shown here:

$$p_i = q_i(GC_{pi})$$
$$p_i = 48.19(\pm0.18) = \pm8.67 \text{ psf}$$

The parapet pressures are calculated using Equation 2.4

$$P_p = q_h(GC_{pa})$$

See Figures 2.5 and 2.6 for pressure distributions (Figures 2.7 and 2.8).

2.3 HORIZONTAL SEISMIC LOADS (CHAPTERS II AND 12 OF ASCE 7)

The earth's outer shell or crust, which encompasses the whole earth, is broken up into frag-mented sections known as plates. The major plates, which are very large, have smaller plates fitting in between, making a network of interlocking sections. The plates vary in thickness, which affect their behavior. These plates move independently to each other by sliding on a viscous layer beneath. The movement of the plates, known as plate tectonics, can be away

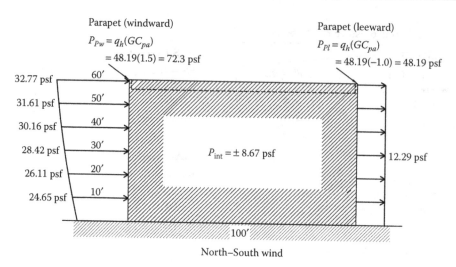

Figure 2.5 Pressure distribution of exterior and interior pressures for north–south wind direction (positive pressure is always toward the surface it is acting on).

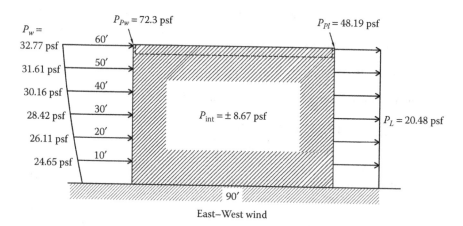

Figure 2.6 Pressure distribution of exterior and interior pressures for east–west wind direction (positive pressure is always toward the surface it is acting on).

or toward each other and they can slip over or under one another. This movement, which can cause plates to collide, can build up energy over time and it is not until the rock material of the plates fails and releases the built-up energy, is the ground motion felt at the earth's surface in the form of oscillating movement. It is this oscillating movement that we design buildings to respond to.

Structural damage due to an earthquake is a function of the characteristics of the earthquake and the site as well as the structural composition of the building. The peak ground acceleration is one indicator of the intensity of the earthquake however the duration of the seismic event and the duration of strong shaking (high peak levels) will contribute to the damage sustained during an earthquake. The site characteristics are extremely important as well when considering damage to a structure. That is, the distance from the origin at the surface/epicenter of the earthquake and the site: the closer to the epicenter the stronger

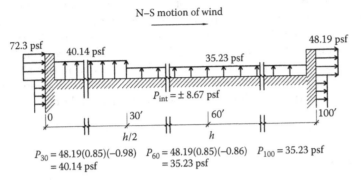

Figure 2.7 Pressure distribution on windward and leeward parapets for north–south wind direction and corresponding roof pressure as per pressure coefficients calculated in step 5.

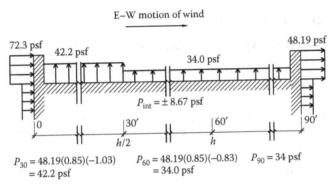

Figure 2.8 Pressure distribution on windward and leeward parapets for east–west wind direction and corresponding roof pressure as per pressure coefficients calculated in step 5.

the ground motion will be felt. The type of geology between the epicenter and the site will contribute to how well the ground motion is conveyed to the site of the building. Ground motion does not travel well through dense material like dense rock and travels extremely well through loose material such as sand. The structural damage sustained to a building will also depend greatly on the age, the construction type and methods used to construct the building. A building, which has been specifically detailed to mitigate risk and damage associated with the energy of an earthquake, will fare much better than an older building that predates structural seismic design provisions.

In order to design a building to sustain the forces imparted by an earthquake, it is necessary to determine the expected magnitude of the earthquake. Building codes have developed geographic distribution of seismic risk, whereby mapped ground motion values associated with the maximum considered earthquake for a region can be readily obtained, see ASCE 7 Chapter 22 for seismic ground motion long-period transition and risk coefficient maps, and used for building design purposes.

The seismic design, for buildings, is based on either a dynamic or a static analysis approach. A dynamic analysis is required when a building is particularly irregular in shape as defined by horizontal and vertical irregularities presented in ASCE 7 Chapter 12. A dynamic analysis will require the building to be mathematically modeled and analyzed. This is typically performed using a proprietary industry acceptable software package, which can model the building's story stiffness and rigidities by using a series of lumped masses, system damping and spring stiffness. A specific design earthquake must be used as the forcing function to generate the building's mode shapes and member forces. The more typical static analysis approach, for seismic design, presented by ASCE 7, is discussed in this chapter. The pertinent seismic design criteria for buildings subject to earthquake ground motion, required for the design and construction of buildings is presented in Chapter 11 of the ASCE 7 standard and summarized here.

2.3.1 Site class

There are six possible site soil classifications: A, B, C, D, E, or F, which is based on the site soil properties. Chapter 20 of the ASCE 7 standard discusses the requirements of site classification. Where sufficient geotechnical detail, regarding the site, is not available to determine site classification, as per Chapter 20, the site class must be classified as D.

2.3.2 Seismic ground motion values

2.3.2.1 Mapped spectral response accelerations

The parameters S_s and S_1 shall be determined from the 0.2 and 1 s spectral response accelerations as shown on the maps noted as follows:

S_s = 0.2 s spectral response acceleration (ASCE 7 Figures 22.1, 22.3, 22.5 and 22.6)
S_1 = 1 s spectral response acceleration (ASCE 7 Figures 22.2, 22.4, 22.5 and 22.6)

2.3.2.2 Site coefficients

F_a = Table 2.5 (ASCE 7 T 11.4-1)
F_v = Table 2.6 (ASCE T 11.4-2)

2.3.2.3 Site coefficients and risk targeted maximum considered earthquake spectral response acceleration parameters

$$S_{MS} = F_a S_s \qquad\qquad (2.5) \text{ (ASCE 7 Equation 11.4-1)}$$
$$S_{M1} = F_v S_1 \qquad\qquad (2.6) \text{ (ASCE 7 Equation 11.4-2)}$$

Table 2.5 Site coefficient, F_a

| Site class | Mapped maximum considered earthquake spectral response acceleration parameter at short period | | | | |
	$S_s \le 0.25$	$S_s = 0.5$	$S_s = 0.75$	$S_s = 1.0$	$S_s \ge 1.25$
A	0.8	0.8	0.8	0.8	0.8
B	1.0	1.0	1.0	1.0	1.0
C	1.2	1.2	1.1	1.0	1.0
D	1.6	1.4	1.2	1.1	1.0
E	2.5	1.7	1.2	0.9	0.9
F			See Section 11.4.7		

Table 2.6 Site coefficient, F_v

Site class	Mapped maximum considered earthquake spectral response acceleration parameter at 1-s period				
	$S_1 \leq 0.1$	$S_1 = 0.2$	$S_1 = 0.3$	$S_1 = 0.4$	$S_1 \geq 0.5$
A	0.8	0.8	0.8	0.8	0.8
B	1.0	1.0	1.0	1.0	1.0
C	1.7	1.6	1.5	1.4	1.3
D	2.4	2.0	1.8	1.6	1.5
E	3.5	3.2	2.8	2.4	2.4
F			See Section 11.4.7		

2.3.2.4 Design spectral acceleration parameters

The design short period spectral accelerations S_{DS} (g) and the design one-second spectral accelerations S_{D1} (g) are obtained as follows:

$$S_{DS} = \frac{2}{3} S_{MS} \qquad\qquad \text{(2.7) (ASCE 7 Equation 11.4-3)}$$

$$S_{D1} = \frac{2}{3} S_{M1} \qquad\qquad \text{(2.8) (ASCE 7 Equation 11.4-4)}$$

2.3.3 Seismic design category

The seismic design category (SDC) dictates the types of structural systems permitted, that is, moment frame, shear wall, braced frame, and so on, the chosen system's permitted height, specific detailing requirements, as well as the type of analysis required. The SDC is a function of the risk category (Table 1.5-1, ASCE 7) of the structure, design spectral accelerations, and as defined as follows:

When $S_1 \geq 0.75$ g and risk category I, II, or III structures are classified as SDC E and risk category IV structures are classified as SDC F.

All other structures are assigned to an SDC based on their risk category and the design spectral response acceleration parameters. Each building and structure shall be assigned to the more severe SDC in accordance with Table 2.7 (ASCE 7 T 11.6-1) or Table 2.8 (ASCE 7 T 11.6-2).

2.3.4 Fundamental period

The fundamental period T(s) of the building is determined in accordance with Section 12.8.2 of ASCE 7 and is usually obtained from a computer model. However, the value for T must not exceed an upper bound that is a function of the approximate period T_a calculated using Equation 2.9.

$$T = \text{computer value} \leq C_u T_a$$

$$C_u = \text{(ASCE 7 T 12.8-1)}$$

Table 2.7 Seismic design category based on short period response acceleration parameter

	Risk category	
Value of S_{DS}	I, II, or III	IV
$S_{DS} < 0.167$	A	A
$0.167 \leq S_{DS} < 0.33$	B	C
$0.33 \leq S_{DS} < 0.50$	C	D
$0.50 \leq S_{DS}$	D	D

Table 2.8 Seismic design category based on 1-s period response acceleration parameter

	Risk category	
Value of S_{DI}	I, II, or III	IV
$S_{DL} < 0.067$	A	A
$0.067 \leq S_{DI} < 0.133$	B	C
$0.133 \leq S_{DI} < 0.20$	C	D
$0.20 \leq S_{DL}$	D	D

2.3.4.1 *Approximate fundamental period*

$$T_a = C_t h_n^x \qquad\qquad (2.9) \ (\text{ASCE 7 Equation 12.8-7})$$

h_n = highest level (ft)
C_t = (ASCE 7 T 12.8-2)
x = (ASCE 7 T 12.8-2)

2.3.5 The equivalent lateral force procedure

2.3.5.1 *Base shear*

The seismic base shear $V(k)$ for a given direction of the building is a function of the effective seismic weight $W(k)$ and is determined as follows:

$$V = C_s W \qquad\qquad (2.10) \ (\text{ASCE 7 Equation 12.8-1})$$

2.3.5.2 *Seismic response coefficient*

The seismic response coefficient, C_s, for the building is determined as follows:

$$C_s = \frac{S_{DS}}{(R/I_e)} \qquad\qquad (2.11) \ (\text{ASCE 7 Equation 12.8-2})$$

where:
   R is the response modification factor (ASCE 7 T 12.2-1)
   I_e is the importance factor (ASCE 7 T 1.5-2)

The value of C_s should not exceed the following:

$$C_s = \frac{S_{D1}}{T(R/I_e)} \text{ for } T \leq T_L \qquad\qquad (2.12) \text{ (ASCE 7 Equation 12.8-3)}$$

$$C_s = \frac{S_{D1}T_L}{T^2(R/I_e)} \text{ for } T \leq T_L \qquad\qquad (2.13) \text{ (ASCE 7 Equation 12.8-4)}$$

And the value of C_s should not be less than the following:

$$C_s = 0.044S_{DS}I_e \geq 0.01 \qquad\qquad (2.14) \text{ (ASCE 7 Equation 12.8-5)}$$

In addition, for structures located where S_1 is equal to or greater than 0.6g, C_s should not be less than the following:

$$C_s = \frac{0.5S_1}{(R/I_e)} \qquad\qquad (2.15) \text{ (ASCE 7 Equation 12.8-6)}$$

T_L is the mapped long-period transition period obtained from ASCE 7 Figure 22-12.

Example 2.2: Design spectral accelerations and SDC

Determine the design spectral accelerations and SDC for a proposed hospital building located in Las Vegas, NV. Assume site class D.

According to ASCE 7 T 1.5-1, *Risk Category of Buildings and Other Structures, for Flood, Wind, Snow, Earthquake, and Ice Loads*, a hospital is an essential facility and is assigned to risk category IV.

The spectral response coefficients S_s and S_1 are obtained from ASCE 7 Maps 22-1 and 22-2, respectively, and they are as follows:

$$S_s = 0.5g \text{ and } S_1 = 0.2g$$

Site coefficients, F_a and F_v, are obtained from Tables 2.5 and 2.6 as follows:

From Table 2.5, for a building with a site class D and $S_s = 0.5g$, $F_a = 1.4$
From Table 2.6, for a building with a site class D and $S_1 = 0.2g$, $F_v = 2.0$

Then,

$$S_{MS} = F_a S_s = (1.4)(0.5g) = 0.7g \qquad\qquad (2.5)$$

$$S_{M1} = F_v S_1 = (2.0)(0.2g) = 0.4g \qquad\qquad (2.6)$$

Hence, the *design spectral accelerations* are calculated as follows:

$$S_{DS} = \frac{2}{3} S_{MS} = \frac{2}{3}(0.7g) = 0.47g \qquad\qquad (2.7)$$

$$S_{D1} = \frac{2}{3} S_{M1} = \frac{2}{3}(0.4g) = 0.27g \qquad\qquad (2.8)$$

The SDC is determined by obtaining the more severe category classification obtained from Tables 2.7 and 2.8.

From Table 2.7, for a building with a risk category IV and $S_{DS} = 0.47g$, SDC = D
From Table 2.8, for a building with a risk category IV and $S_{D1} = 0.27g$, SDC = D
Hence, SDC is D.

Example 2.3: Seismic base shear

Determine the base shear for the proposed hospital building discussed in Example 2.2, assume the seismic-force-resisting systems is a five-story steel special moment frame (SSMF) shown below, see Figure 2.9.

A building assigned to risk category IV, such as a hospital will have an earthquake importance factor $(I_e) = 1.5$, as per ASCE 7 T 1.5-2, *Importance Factors by Risk Category of Buildings and Other Structures for Snow, Ice, and Earthquake Loads.*

The response modification factor R is obtained from ASCE 7 T 12.2-1 *Design Coefficients and Factors for Seismic-Force-Resisting Systems.* From Section C, moment-resisting frame systems, an SSMF is permitted to be used in SDC D and has an R = 8.

In order to calculate the base shear for the building, the period of the structure is needed, along with the seismic response coefficient C_s.

Step 1: Determine the period of the structure:
   The approximate fundamental period, T_a, can be used.

$$T_a = C_t h_n^x \tag{2.9}$$

The height of the steel frame $h_n = 55.5$ ft.
   For an SSMF, obtain C_t and x from ASCE 7 T 12.8-2:

$C_t = 0.028$
$x = 0.8$

Then, $T_a = C_t h_n^x = (0.028)(55)^{0.8} = 0.691$ s
$T_L = 8$ s (ASCE 7 Figure 22-12)
Step 2: Determine the seismic response coefficient:
   Using S_{DS} and S_{D1} as determined in the previous example, the design value of C_s is the smaller value of

$$C_s = \frac{S_{DS}}{(R/I_e)} = \frac{0.47}{(8/1.5)} = 0.0881 \tag{2.11}$$

and

$$C_s = \frac{S_{D1}}{T(R/I_e)} = \frac{0.27}{(0.691)(8/1.5)} = 0.0733 \text{ for } T \le T_L \tag{2.12}$$

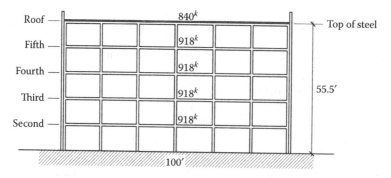

Figure 2.9 Five-story steel special moment frame and supported roof and floor weights.

$$C_s = \frac{S_{D1}T_L}{T^2(R/I_e)} = \frac{(0.27)(8)}{(0.691)^2(8/1.5)} = 0.8482 \text{ for } T > T_L \tag{2.13}$$

but C_s must not be less than

$$C_s = 0.044S_{DS}I_e = (0.044)(0.47)(1.5) = 0.0310 \geq 0.01 \tag{2.14}$$

Since S_1 is not equal or greater than 0.6g, Equation 2.16 does not apply.
Therefore, the design value of $C_s = 0.0733$.
Step 3: Determine the seismic base shear:
The seismic base shear is determined by

$$V = C_sW \tag{2.15}$$

The effective seismic weight W is generally considered to be the dead load of the building above the base. From Figure 2.7, $W = 4.512$ kips.
Then, the seismic base shear is

$$V = (0.0733)(4512) = 331 \text{ kips}$$

Example 2.4: Seismic lateral load on parapet

Calculate the seismic design load for the parapet shown in Figure 2.10, which is attached to the proposed hospital building discussed in the previous examples.

The parapet is a cantilevered structure braced below its center of mass by the roof framing system.

Since a parapet is considered an architectural component, the design seismic load of a cantilever parapet is determined using the specifications of ASCE 7 Chapter 13, *Seismic Design Requirements for Nonstructural Components*.

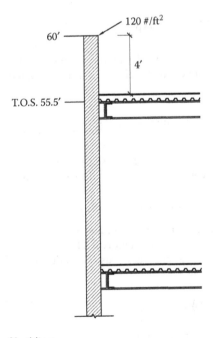

Figure 2.10 Upper wall section of building.

The horizontal seismic design force, F_p, is determined in accordance with the following equation and is applied at the component's center of gravity:

$$F_p = \frac{0.4a_p S_{DS} w_p}{(R_p/I_p)}\left(1+2\frac{z}{h}\right)$$ (2.16) (ASCE 7 Equation 13.3-1)

The component importance factor, I_p, is either 1.0 or 1.5 when the component is in or attached to a risk IV category structure, see ASCE 7 Section 13.1.3. The parapet, under consideration, is part of a proposed hospital building assigned to risk category IV, $I_p = 1.5$.
   Use ASCE 7 T 13.5-1, *Coefficients for Architectural Components* to obtain a_p and R_p.
   Then, for cantilever elements (braced to structural frame below its center of mass), parapets: $a_p = 2.5$, $R_p = 2.5$
   z = height of the point of attachment of the component to the structure = 56 ft
   h = average roof height of the structure = 56 ft
   Therefore, using $S_{DS} = 0.47g$, as previously calculated in Example 2.2, F_p can be determined as calculated here.

$$F_p = \frac{0.4(2.5)(0.47)120 \text{ psf}}{(2.5/1.5)}\left(1+2\frac{56}{56}\right) = (33.84)(3) = 101.52 \text{ psf}$$

F_p = is not required to be taken as greater than

$$F_{p,max} = 1.6 \, S_{DS} I_p w_p$$ (2.17) (ASCE 7 Equation 13.3-2)

$$= 1.6(0.47)(1.5)120 = 135.36 \text{ psf}$$

and F_p is not to be taken less than

$$F_{p,min} = 0.3 \, S_{DS} I_p w_p$$ (2.18) (ASCE 7 Equation 13.3-3)

$$= 0.3(0.47)(1.5)120 = 25.38 \text{ psf}$$

Thus, $F_{p,min} < F_p < F_{p,max}$, the seismic design force value is okay.
   Thus, the horizontal seismic force on the parapet is calculated in a per linear foot basis as follows:

$$F_p = \left(101.52\frac{\#}{\text{ft}^2}\right)(4' \text{ tall parapet}) = 406.8 \, \frac{\#}{\text{linear ft of wall}}$$

It is important to note that for vertically cantilevered systems, the component must also be designed for a concurrent vertical force equal to $\pm 0.2 S_{DS} W_p$.

Example 2.5: Out-of-plane seismic load on walls

Determine the out-of-plane seismic load and anchorage force on the fifth floor and roof level; see the building section shown in Figure 2.11.
   Structural walls and their anchorages must be designed for a force normal to their surface equal to the following:

$$F_p = 0.4 S_{DS} I_e W_{wall} \geq 0.1 W_{wall}$$ (2.19) (ASCE 7 Section 12.11.1)

where W_{wall} = weight of the structural wall.

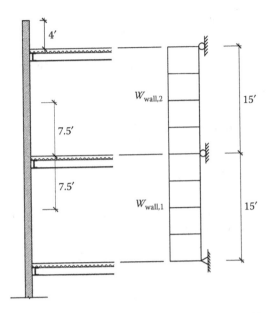

Figure 2.11 Upper wall section of building and free-body diagram of horizontal loading.

Using $S_{DS} = 0.47g$, $I_e = 1.5$ and $W_{wall} = 120(\#/ft^2)$ as previously determined, the horizontal seismic force, F_p, is calculated as follows:

$$F_p = (0.4)(0.47)(1.5)(120) = 33.84\frac{\#}{ft^2}$$

which is greater than $0.1W_{wall}$, therefore the value of F_p is OK.

The wall is connected at the floor diaphragms and the roof. The tributary width of wall at the fifth floor and roof is 15 and 7.5 ft, respectively, therefore the horizontal forces at the fifth floor and the roof is $(33.84)(15) = 507.6(\#/ft)$ and similarly $253.8(\#/ft)$.

In order to calculate the wall anchorage forces, ASCE 7 Section 12.11.2, *Anchorage of Structural Walls and Transfer of Design Forces into Diaphragms*, specifies the following equation for F_p as the force, the connection between the wall and the supporting construction must resist.

$$F_p = 0.4S_{DS}k_a I_e W_p \qquad\qquad (2.20)\ (ASCE\ 7\ Equation\ 12.11\text{-}1)$$

Buildings with a floor construction consisting of concrete filled metal deck with span-to-depth ratios of 3 or less and having no horizontal irregularities are, permitted to be, idealized as rigid diaphragm buildings as per ASCE Section 12.3.1.2, and is the case with the building presented in this example.

Assuming a rigid diaphragm condition allows Equation 2.20 to be reduced to

$$F_p = 0.4S_{DS}I_e W_p$$

because k_a is permitted to be taken as 1.0 for rigid diaphragm structures.

W_p = to the weight of the wall tributary to the anchor at the fifth floor

$$\text{which is} = \left(120\frac{\#}{ft^2}\right)(15 \text{ ft}) = 1800\frac{\#}{\text{foot of wall}}$$

Therefore, to determine the out-of-plane anchorage force at the roof and the floor diaphragms
  Anchorage forces:

$$F_{roof} = (0.4)(0.47)(1.5)(1800/2) = 253.83\frac{\#}{ft}$$

$$F_{5th \text{ floor}} = (0.4)(0.47)(1.5)(1800) = 507.6\frac{\#}{ft}$$

which are the same values as the out-of-plane forces previously calculated. However, wall anchorage forces cannot be taken less than

$$F_{p(min)} = 0.2k_aI_eW_p$$

with k_a taken as 1.0,

$$F_{p(min)roof} = (0.2)(1.5)\left(\frac{1800}{2}\right) = 270\frac{\#}{ft}$$

$$F_{p(min)5th \text{ floor}} = (0.2)(1.5)(1800) = 540\frac{\#}{ft}$$

Therefore, the design values for the roof and the fifth floor are 270(#/ft) and 540(#/ft), respectively.

2.4 VERTICAL SEISMIC LOAD EFFECT

Seismic load effects are the axial, shear and flexural member forces resulting from the application of horizontal and vertical seismic forces. The seismic load effect E and the governing load combinations including E are given in ASCE 7 Section 12.4.2 and they are

$$E = E_h + E_v \qquad\qquad \text{(2.21) (ASCE 7 Equation 12.4-1)}$$

used with the strength design load combination 5 in ASCE Section 2.3.2

$$1.2D + 1.0E + L + 0.2S, \text{ and}$$

$$E = E_h - E_v \qquad\qquad \text{(2.22) (ASCE 7 Equation 12.4-2)}$$

used with the strength design load combination 7 in ASCE Section 2.3.2 (where the negative sign is the upward action)

$$0.9D + 1.0E$$

The horizontal seismic load effect,

$$E_h = \rho Q_E \qquad\qquad \text{(2.23) (ASCE 7 Equation 12.4-3)}$$

where:

Q_E is horizontal seismic forces, and

ρ is the redundancy factor, which will be covered in the following chapter of this book.

The vertical seismic load effect,

$$E_v = 0.2S_{DS}D \qquad\qquad \text{(2.24) (ASCE 7 Equation 12.4-4)}$$

Example 2.6: Upward seismic forces on a beam

Determine the vertical seismic force for the cantilever beam shown in Figure 2.12 for strength design load cases, $S_{DS} = 0.47g$, $w_{\text{self weight}} = 500$ 3/ft.

The design of horizontal beams in an SDC D region must consider the effects due to E, which means the load combinations involving seismic must be considered (Figure 2.13). Substituting $E = E_h + E_v$ in the load case 5) $1.2D + 1.0E + L + 0.2S$ gives: $1.2D + 1.0(E_h + E_v) + L + 0.2S$, which becomes $1.2D + 0.2S_{DS}D + \rho Q_E + L + 0.2S$, and because there is no horizontal seismic force ($Q_E = 0$) and obviously no live load or snow load, the load combination reduces down to

5) $(1.2 + 0.2S_{DS})D$.

Similarly, substituting $E = E_h - E_v$ into load case $0.9D + 1.0E$ is

7) $(0.9 - 0.2S_{DS})D$.

These load cases are the seismic load combination for strength design given in ASCE 7 Section 12.4.2.3, as shown here,

5) $(1.2 + 0.2S_{DS})\, D + \rho Q_E + L + 0.2S$

7) $(0.9 - 0.2S_{DS})D + \rho Q_E$

Then, the governing load combination including the downward seismic effect is

$$U_5 = (1.2 + 0.2S_{DS})\, D = [1.2 + 0.2(0.47)]500 = 647\ \#/\text{ft}$$
$$U_7 = (0.9 - 0.2S_{DS})\, D = [0.9 - 0.2(0.47)]500 = 503\ \#/\text{ft}$$

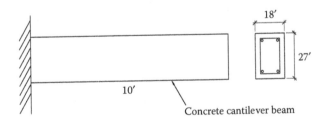

Figure 2.12 Cantilever beam, 10 ft long, with a self weight of 500 #/ft.

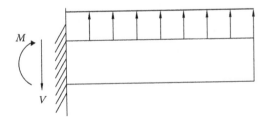

Figure 2.13 Minimum uniform upward force on cantilever beam.

Consequently, there is not any net upward load on the beam and load case 5 produces the maximum downward load on the beam.

However, a provision exists in ASCE 7 (Section 12.4.4) specifically requiring a minimum upward design force for horizontal cantilevers for seismic design categories D, E, and F, of 0.2 times the dead load, in addition to applicable load combinations for the seismic design.

Therefore, the minimum upward force, which must be considered for the design of the horizontal beam, is

$$W_u = -0.2(500) = -100 \text{ #/ft}$$

$$V = 100(10) = 1000 \text{ #}$$

$$M = \frac{(100)(10)^2}{2} = 5000 \text{ft} - \text{#}$$

Chapter 3

Lateral force distribution

3.1 WALL RIGIDITIES

The rigidity of a wall, often called the stiffness, will determine how well a wall will react to lateral forces, that is, the amount the wall will deflect when loaded laterally. The rigidity of a wall is dependent on the cross-sectional dimensions, the height of the wall, and the material it is made of—specifically how well it behaves in bending and in shear, and the connection at the top and bottom of the wall. A wall, which is fixed securely at its bottom and is free to rotate at the top of the wall, is considered a cantilever wall, which is similar in behavior to a cantilever beam. A wall, which is restrained at the bottom and top against rotation, is considered a fixed wall, which is similar to a beam fixed at both ends.

The rigidity of a wall is defined as the reciprocal of the deflection. For example, for wall i the rigidity $R_i = 1/\delta_i$.

3.1.1 Cantilever wall

For a wall, which is fixed at its base or foundation and is a vertical cantilever, the deflection is (Figure 3.1):

$$\delta_{cant} = \delta_m + \delta_v = \frac{Ph^3}{3E_mI} + \frac{1.2Ph}{AE_v} \tag{3.1}$$

where:
δ_m is the deflection due to flexural bending
δ_v is the deflection due to shear
P is the lateral force applied to the wall
h is the height of the wall
A is the cross-sectional area of the wall
I is the cross-sectional moment of inertia of the wall in the direction of bending
E_m is the modulus of elasticity in compression
$E_v = G$ is the modulus of elasticity in shear ($E_v = 0.4E_m$)

3.1.2 Fixed wall

For a wall, which is fixed both at its base and at the top and restrained against rotation, the deflection is as follows (Figure 3.2):

$$\delta_{fixed} = \delta_m + \delta_v = \frac{Ph^3}{12E_mI} + \frac{1.2Ph}{AE_v} \tag{3.2}$$

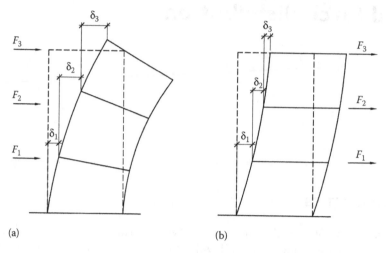

(a)   (b)

Figure 3.1 Shear wall deformations for a multilevel cantilever wall: (a) flexural deformation and (b) shear deformation.

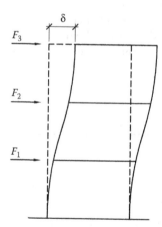

Figure 3.2 Shear wall deformations for a multilevel fixed wall.

In practice, it is often sufficient to use relative rigidities for the wall of the building being designed. The deflection equations can be simplified by substituting $I = (1/12)bL^3$, $A = bL$, $P = 1$, $E_v = 0.4E_m$, and $E_m = 1$
where:

    b is the thickness or width of the wall
    L is the length of the wall in the direction of force

Then Equations 3.1 and 3.2 can be written as follows:

$$\delta_{cant} = \delta_m + \delta_v = \frac{4}{b}\left(\frac{h}{L}\right)^3 + \frac{3h}{bL} \tag{3.3}$$

$$\delta_{fixed} = \delta_m + \delta_v = \frac{1}{b}\left(\frac{h}{L}\right)^3 + \frac{3h}{bL} \tag{3.4}$$

If the walls are all of the same thickness, the width b can be set to 1 and the deflection equations can be further simplified.

Then it follows that the relative rigidities of the walls are obtained by using the reciprocal of the simplified deflection equation, $R_i = 1/\delta_{\text{cant}}$ and $R_i = 1/\delta_{\text{fixed}}$.

3.2 RELATIVE RIGIDITY FORCE DISTRIBUTION (RIGID DIAPHRAGM ANALYSIS)

Floors and flat roofs of buildings are considered horizontal diaphragms, which are designed to distribute lateral loads to the vertical supporting structural elements of the lateral load resisting system such as shear walls or frames. A diaphragm is considered to be *rigid* or *flexible* based on its stiffness. A diaphragm, which is idealized as rigid, is typically a solid concrete slab or concrete on metal deck slab having a span-to-depth (aspect) ratio of 3 or less. Flexible diaphragms are typically cold-formed metal joists with plywood decking, which when loaded laterally behaves much the same way a beam would behave while bending, thereby distributing loads based on the tributary widths between adjacent vertical lateral load resisting elements.

A rigid diaphragm analysis assumes that the diaphragm is sufficiently stiff to retain its form and engage the vertical elements based on their stiffness or rigidity. That is, in the case when vertical elements such as shear walls or frames have different rigidities, the stiffer elements will attract a greater load due to the resistance they offer. A rigid diaphragm analysis will determine the maximum shear force distributed to each of the vertical elements, which support the diaphragm.

3.2.1 Center of mass

The center of mass needs to be calculated to determine the location of the seismic shear to be applied to the building. All of the masses, which are tributary to the diaphragm, including the mass of the diaphragm itself are considered in determining the center of mass of the floor under consideration.

Example 3.1: Locating the center of mass of a building

A plan view of a typical floor of a multistory building is shown in Figure 3.3. Concrete shear walls resist lateral forces in both directions. Determine the location of the center of mass (CM) at the given floor level. The mass of the floor diaphragm is uniformly distributed and the weights of the concrete shear walls are as noted in the figure and are tributary to the diaphragm—that is, half of the wall height below and half of the wall height above the diaphragm constitutes the weight of the wall for analysis. The noncantilevered portion of the floor diaphragm is 85 psf and the cantilevered portion of the diaphragm is 65 psf. The floors of the building are 6 in. thick concrete slab on metal deck and are idealized as a rigid diaphragm. The floor-to-floor height is 12 ft.

The center of mass coordinates can be determined by the following equations:

$$x_{\text{cm}} = \frac{\sum w_{i,y} x_i}{w_{i,y}} \tag{3.5}$$

$$y_{\text{cm}} = \frac{\sum w_{i,x} y_i}{w_{i,x}} \tag{3.6}$$

The weight and center of gravity for each building component with respect to the x- and y-directions are presented in Table 3.1.

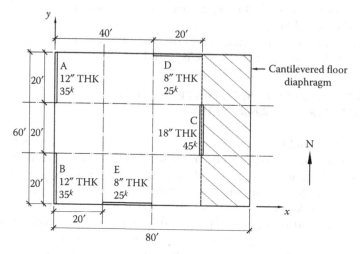

Figure 3.3 Plan view of the building, showing the walls and extent of typical floor diaphragm.

Table 3.1 Tabulated weights and center of gravity for each building component

Building component	Weight W_i (k)	X_i (ft)	y_i (ft)	W_iX_i (ft-k)	W_iY_i (ft-k)
Wall					
A	35	0.5	50	17.5	1750
B	35	0.5	10	17.5	350
C	45	60	30	2700	1350
D	25	50	59.67'	1250	1491.75
E	25	30	$\frac{4}{12} = 0.333'$	750	8.33
Diaphragm					
Noncantilevered	$\frac{60 \times 60 \times 85}{1000} = 306^k$	30'	30'	9180	9180
Cantilevered	$\frac{20 \times 60 \times 85}{1000} = 78^k$	70'	30'	5460	2340
	549			19375	16470

Then the coordinates for the center of mass can be readily determined as follows:

$$x_{cm} = \frac{\sum w_{i,y} x_i}{w_{i,y}} = \frac{19,375}{549} = 35.3$$

$$y_{cm} = \frac{\sum w_{i,x} y_i}{w_{i,x}} = \frac{16,470}{549} = 30$$

The location of the center of mass of the building is shown in Figure 3.4.

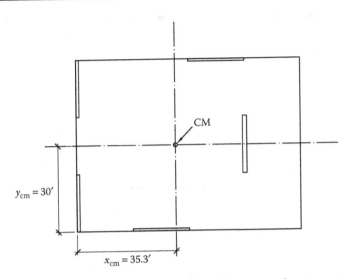

Figure 3.4 Plan view of the building, illustrating the location of the center of mass.

3.2.2 Center of rigidity

The center of rigidity is the origin whereby the seismic force applied to the center of mass of the floor will rotate. Each vertical element, that supports the floor diaphragm laterally, the wall or frame, which is tributary to the diaphragm, has a specific rigidity and needs to be determined in order to calculate the center of rigidity of the floor. When forces are parallel to the length of the wall, its rigidity must be determined. When forces are perpendicular to the wall, its rigidity is conservatively taken as zero.

Example 3.2: Locating the center of rigidity

For the building discussed in Example 3.1, determine the location of the center of rigidity (CR) and the value of the torsional rigidity, J, for the lateral force resisting system configuration as shown.

The walls of the multistory building in Example 3.1 have a floor slab at the bottom and at the top of the wall at each level, which prevents rotation of the top of the wall and creates a *fixed* wall condition. Using deflection Equation 3.4 for a fixed wall, as previously discussed, the rigidity of each of the walls is determined as given in Table 3.2.

$$\delta_{fixed} = \delta_m + \delta_v = \frac{1}{b}\left(\frac{h}{L}\right)^3 + \frac{3h}{bL} \tag{3.4}$$

Table 3.2 Tabulated wall deflections and rigidities

Wall	Thickness b (in.)	Height h (ft)	Length L (ft)	Deflection δ	Rigidity
A	12	12	20	0.168	5.95
B	12	12	20	0.168	5.95
C	18	12	20	0.112	8.93
D	10	12	20	0.2016	4.96
E	10	12	20	0.2016	4.96

The center of rigidity coordinates can be determined by the following equations:

$$x_{cr} = \frac{\sum R_{i,y} x_i}{R_{i,y}} \qquad\qquad (3.7)$$

$$y_{cr} = \frac{\sum R_{i,x} y_i}{R_{i,x}} \qquad\qquad (3.8)$$

The wall rigidities for each wall with respect to their direction of resistance are given in Table 3.3 and Figure 3.5.

Then the coordinates for the center of rigidity is determined as follows (Figure 3.6):

$$x_{cr} = \frac{\sum R_{i,y} x_i}{R_{i,y}} = \frac{541.76}{20.83} = 26 \text{ ft}$$

$$y_{cr} = \frac{\sum R_{i,x} y_i}{R_{i,x}} = \frac{297.60}{9.92} = 30 \text{ ft}$$

Table 3.3 Tabulated wall rigidities

Wall	R_{iy}	R_{ix}	X_i (ft)	Y_i (ft)	$R_{iy}X_i$	$R_{ix}y_i$
A	5.95	0	0.5'		2.98	
B	5.95	0	0.5'		2.98	
C	8.93	0	60'		535.8	
D	0	4.96		0.33'		1.64
E	0	4.96		59.67'		259.96
	20.83	9.92			541.76	297.60

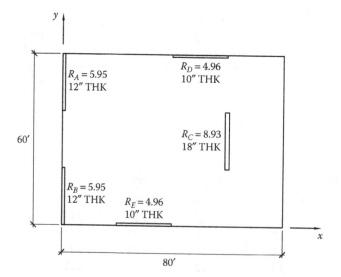

Figure 3.5 Plan view of the building, showing the walls and corresponding wall rigidities.

3.2.3 Polar moment of inertia

The seismic force produced by ground acceleration, which excites the building's masses, acts at the center of mass of the building. Because the center of mass and center of rigidity is not usually in the same location, a torsional force or moment is produced at the center of rigidity of the building. The response of the building to this torsional force is predicted by the rotational moment of inertia or the *polar moment of inertia*, which is based on the layout and rigidity of the walls.

The polar moment of inertia is generated around the coordinates of the center of rigidity. Each wall's distance from the center of rigidity is calculated by

$$\overline{x}_i = x_i - x_{cr} \tag{3.9}$$

$$\overline{y}_i = y_i - y_{cr} \tag{3.10}$$

Then, the polar moment of inertia, J, is obtained by the following equation:

$$J = \sum R_{i,y}\overline{x}_i^2 + \sum R_{i,x}\overline{y}_i^2 \tag{3.11}$$

The distance of each wall from the center of rigidity of the building is calculated below and shown in Table 3.4.

Wall A: $\overline{x} = 0.5' - 26.0' = -25.5'$
Wall B: $\overline{x} = 0.5' - 26.0' = -25.5'$
Wall C: $\overline{x} = 60.0' - 26.0' = 34.0'$
Wall D: $\overline{y} = 59.67' - 30.0' = 29.67'$
Wall E: $\overline{y} = 0.33' - 30.0' = -29.67'$

From Table 3.4, the polar moment of inertia is calculated as follows:

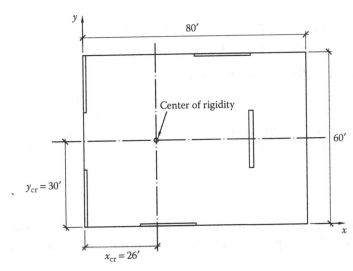

Figure 3.6 Plan view of the building, illustrating location of the center of rigidity.

Table 3.4 Tabulated information for polar moment of inertia

Wall	R_{iy}	R_{ix}	$\bar{x}_i$ (ft)	$\bar{y}_i$ (ft)	$R_{iy}\bar{x}_i^2$	$R_{ix}\bar{y}_i^2$
A	5.95	0	−25.5′	–	3869.0	–
B	5.95	0	−25.5′	–	3869.0	–
C	8.93	0	34.0′	–	10323.1	–
D	0	4.96	–	29.67′	–	4366.3
E	0	4.96	–	−29.67′	–	4366.3
					18061.1	8732.6

$$J = \sum R_{i,y}\bar{x}_i^2 + \sum R_{i,x}\bar{y}_i^2 = 18{,}061 + 8{,}732.6 = 26{,}793.7 \text{ ft}^2$$

3.2.4 Eccentricity

According to ASCE 7, Section 12.8.4.2, *accidental torsion* is the moment arm created by the distance between the center of mass and the center of rigidity, for buildings with rigid diaphragms, must be increased by an offset of the center of mass by 5% of the dimension of the structure, perpendicular to the direction of the applied force to account for an accidental eccentricity.

For an applied seismic load in the north-south direction of the building discussed in Example 3.1, the perpendicular dimension is 80 ft, hence the offset to account for accidental eccentricity is $0.05(80) = 4$ ft to the left and right of the center of mass, thereby creating three locations of the center of mass to apply the seismic force, as shown here:

$x_{cm} = 31.3$ ft, $x_{cm} = 35.3$ ft, and $x_{cm} = 39.3$ ft

Hence, the eccentricity is the distance from the center of rigidity (moment arm) as follows:

$e_1 = 31.3 - 26.0 = 5.3$ ft

$e_2 = 35.3 - 26.0 = 9.3$ ft

$e_3 = 39.3 - 26.0 = 13.3$ ft

Each location should be evaluated individually to assess the governing (highest) forces at each wall. See Figure 3.7 for the location of the seismic force at the center of mass of the building.

3.2.5 Wall shears (direct and torsional)

The focus of a rigid diaphragm analysis is to determine the wall shears due to an applied horizontal seismic force. The total shear in a wall is due to direct and torsional shear.

The *direct shear* in each of the walls, due to a shear force V_y illustrated in Figure 3.8, is proportional to its rigidity and is calculated as follows:

$$V_{Di\,y} = V_y \frac{R_{i,y}}{\sum R_{i,y}} \tag{3.12}$$

where $\sum R_{i,y}$ only includes the sum of rigidities of the walls that are parallel to the line of force applied. Walls that are perpendicular to the applied force have a rigidity of zero and are not included.

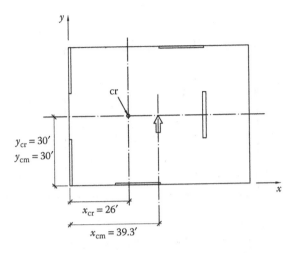

Figure 3.7 Location of the seismic force at the center of mass of the building.

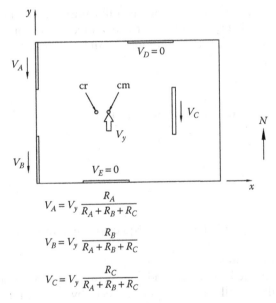

$$V_A = V_y \frac{R_A}{R_A + R_B + R_C}$$

$$V_B = V_y \frac{R_B}{R_A + R_B + R_C}$$

$$V_C = V_y \frac{R_C}{R_A + R_B + R_C}$$

Figure 3.8 Illustration of direct shears with respect to applied seismic force V_y applied at the center of mass.

Similarly for a seismic force applied in the x-direction, we have

$$V_{D,i\,x} = V_y \frac{R_{i,x}}{\sum R_{i,x}} \tag{3.13}$$

The *torsional shears* are due to the eccentric seismic force, applied to the center of mass (see Figure 3.9), and rotating about the center of rigidity. The shear walls resist this torsional force and develop shears.

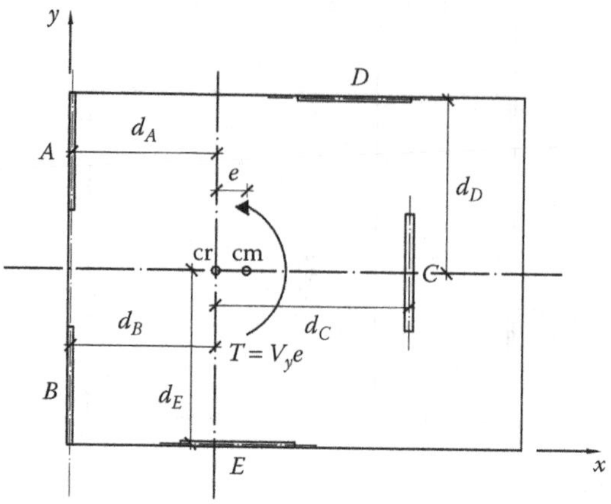

Figure 3.9 Illustration of torsional moment about the center of rigidity with applied seismic force at the center of mass.

The torsional moment $T = V_y(e) = \Sigma V_i d_i$ for equilibrium to exist. From this and the fact that the wall displacement is proportional to the distance away from the center of gravity, the shear at each wall due to the torsional moment can be derived as

$$V_{T,i} = V(e)\frac{R_i d_i}{J} \tag{3.14}$$

which can also be written in terms of x- and y-wall shears for each wall i as follows:

$$V_{T,i\,x} = V_x(e)\frac{R_{i,x}\bar{y}_i}{J} \tag{3.15}$$

$$V_{T,i\,y} = V_y(e)\frac{R_{i,y}\bar{x}_i}{J} \tag{3.16}$$

Then the total shear for each of the walls is equal to the sum of the direct shear and the torsional shear

$V_i = $ direct shear + torsional shear.

For a seismic force applied in the x- and y-directions, the total shear can be written as shown here, for walls parallel to the applied force, by combining Equations 3.13 and 3.15 and Equations 3.12 and 3.16, respectively.

$$V_{i,x} = V_x\frac{R_{i,x}}{\sum R_{i,x}} + V_x(e)\frac{R_{i,x}\bar{y}_i}{J} \tag{3.17}$$

$$V_{i,y} = V_y\frac{R_{i,y}}{\sum R_{i,y}} + V_y(e)\frac{R_{i,y}\bar{x}_i}{J} \tag{3.18}$$

For walls perpendicular to the applied force, there is no direct shear. However, perpendicular walls attract shear forces due to the torsional moment. Hence, for walls perpendicular to the applied seismic force in the x- and y-directions, the shear forces is shown here, respectively.

$$V_{i,x} = 0 + V_y(e)\frac{R_{i,x}\bar{y}_i}{J} \qquad (3.19)$$

$$V_{i,y} = 0 + V_x(e)\frac{R_{i,y}\bar{x}_i}{J} \qquad (3.20)$$

where e includes the accidental torsion and $\bar{x}_i = x_i - x_{cr}$ and $\bar{y}_i = y_i - y_{cr}$.

Example 3.3: Determine the direct and torsional wall shears

For the building discussed in Example 3.1 determine the shear force in walls A, B, C, D, and E for a seismic force $V = 150$ kips applied in the y-direction. $J = 26,793.7$ ft² and $e = 13.3'$.

Solution

The total shear in the walls (A, B, and C) parallel to the applied force is

$$V_{\text{Total}\,i,y} = V_{\text{Direct}\,i,y} + V_{\text{Torsional}\,i,y}$$

$$V_{i,y} = V_y\frac{R_{i,y}}{\sum R_{i,y}} + V_y(e)\frac{R_{i,y}\bar{x}_i}{J}$$

and the total shear in the walls (D and E) perpendicular to the applied force is

$$V_{\text{Total}\,i,x} = 0 + V_{\text{Torsional}\,i,x}$$

$$V_{i,x} = 0 + V_y(e)\frac{R_{i,x}\bar{y}_i}{J}$$

Then, the components of the equations for the direct and torsional shears are calculated in Table 3.5. See Table 3.4 for $\bar{x}_i$ and y_i values.

From Table 3.5, the design shears can be assembled as shown here:

$$V_{A_y} = 42.85 - 11.3 = 31.55 \text{ kips}$$

$$V_{B_y} = 42.85 - 11.3 = 31.55 \text{ kips}$$

$$V_{C_y} = 64.31 + 22.6 = 86.91 \text{ kips}$$

$$V_{D_x} = 0 + 10.96 = 10.96 \text{ kips}$$

$$V_{E_x} = 0 - 10.96 = -10.96 \text{ kips}$$

The design shears for each wall and the direction and location of the applied load are shown in Figure 3.10.

In practice, the seismic shear force is applied in each direction of the building, thereby requiring the horizontal distribution of the seismic force applied in the orthogonal direction.

Table 3.5 Direct and torsional shears for an applied seismic force in the y-direction

			y-direction load, $V_y = 150^k$, $T = V_y e_x$				
Wall	R_{iy}	R_{ix}	V_{Diy} (k)	$\bar{x}_i$ (ft)	$\bar{y}_i$ (ft)	V_{Tiy} (k)	V_{Tix} (k)
A	5.95	–	42.85′	–25.5′	–	–11.3′	–
B	5.95	–	42.85′	–25.5′	–	–11.3′	–
C	8.93	–	64.31′	34.0′	–	22.6	–
D	–	4.96	–	–	29.67′	–	10.96′
E	–	4.96	–	–	–29.67′	–	–10.96′
	20.83	9.92					

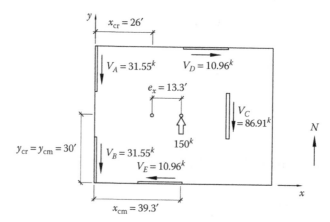

Figure 3.10 Floor plan showing the wall shears and direction.

3.3 FLEXIBLE DIAPHRAGMS

A horizontal flexible diaphragm is similar in behavior to a simply supported beam on its side, that is, the diaphragm is designed to sustain imposed shear and bending forces due to lateral loads imparted uniformly across the edge of the diaphragm. A building with a roof diaphragm and perimeter walls, as shown in Figure 3.11, act as vertical lateral load resisting elements, which provide support to the diaphragm at its ends and allows the diaphragm to bend as if it was simply supported. The field or panel of the diaphragm is similar to the web of a beam and must sustain the shear forces across it. The edges of the diaphragm, called chords are connected to the panel and are designed to sustain the compression and tension forces, which develops in them when the diaphragm bends (see Figure 3.12).

Wind and earthquakes are the typical lateral forces a building must be designed to sustain. The wind force as applied on the building in Figure 3.11 is vertically distributed between the ground level and the roof, that is, half the wind load is applied to the roof diaphragm.

The wind force is applied as a uniform load across the diaphragm's edge and promotes bending as shown in Figure 3.12, thereby developing tension and compression forces in the chords perpendicular to the direction of the force. The chords, which are parallel to the force, act as struts, which *drag* the force, into the shear walls at the end of the diaphragm.

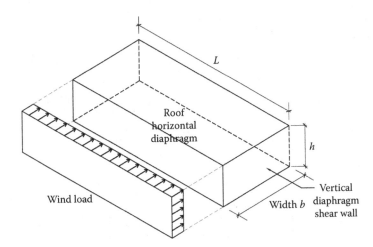

Figure 3.11 Diaphragm building with lateral wind load applied to long side.

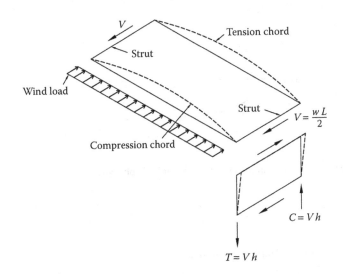

Figure 3.12 Interface of vertical and horizontal diaphragms.

The shear walls respond to the shear at the top of the wall by deforming and translating the shear force to the base of the wall. Overturning forces in the walls need to be addressed in a proper design of the shear walls. The shear force or vertical supports at the ends of the diaphragm are shown in Figure 3.13.

The chord forces (tension and compression) are found by dividing the diaphragm moment (M) by the eccentricity (e) or the width of the diaphragm parallel to the direction of the force, as shown here:

$$T = C = \frac{M}{e} = \frac{wL^2}{8b} \tag{3.21}$$

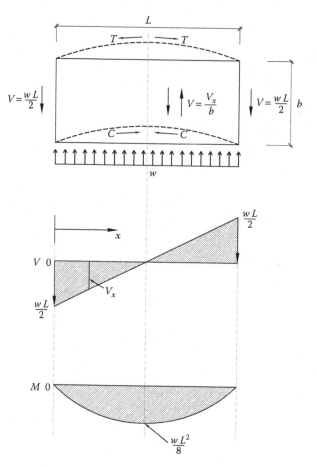

Figure 3.13 Shear force and maximum moment in a flexible diaphragm.

Example 3.4: A flexible diaphragm analysis

A roof plan and elevation of a single-story precast concrete tilt-up building with panel-ized wood roof (flexible roof diaphragm) is shown in Figure 3.14. A wind load of $25(\#/ft^2)$ is applied in the east-west direction (Figure 3.15).

1. Determine the wind force transmitted to the roof diaphragm from the tilt-up wall panels.
2. Determine the chord forces and the shear at the struts of the diaphragm.

Solution

1. Summing moments about A to determine the wind load transmitted to the roof diaphragm

$$+\!\!\curvearrowright \Sigma M_A = 0$$

$$25(25)\frac{25}{2} - B\,(21) = 0$$

$$B = 372^{\#/ft}$$

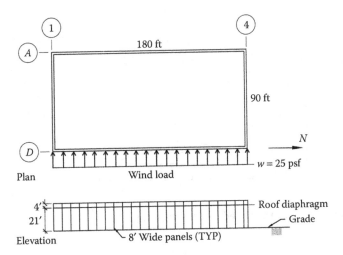

Figure 3.14 Plan and elevation of tilt-up one-story building.

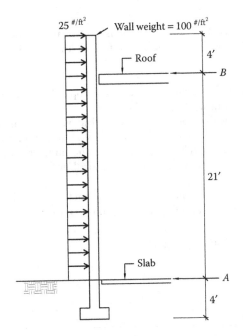

Figure 3.15 Building wall section.

2. The diaphragm moment is calculated as follows (Figure 3.16):

$$M = \frac{wl^2}{8} = \frac{372^{\#/ft}(180')^2}{8} = 1,506,600^{ft\cdot\#}$$

Then, the tension and compression chord forces are as follows:

$$T = C = \frac{M}{b} = \frac{1,506,600^{ft\cdot\#}}{90'} = 16,740^{\#}$$

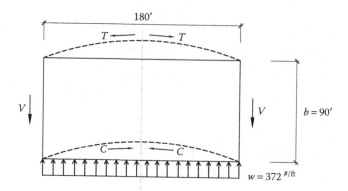

Figure 3.16 The wind load applied to the edge of the diaphragm.

Shear in the struts at end walls along grid lines ① and ④ is as follows:

$$V = \frac{wl}{2} = \frac{372^{\#/ft}(180')}{2} = 33,480^{\#}$$

The building is modified and a shear wall is added along grid line ② as shown in Figure 3.17. The wall does not extend to the full width of the building. Determine the following:

1. The shear at each of the walls parallel to the applied wind load
2. The force in the drag strut along grid line ②
3. Assume that the wall panels are connected at vertical wall joints. The in-plane design strength of the connections $\phi V_n = 4.8k$. Using a load factor of 1.6 for the wind applied to the diaphragm, determine the number of connections required between two adjacent 8′ wide panels along gridline ④

1. For a flexible diaphragm, the wind loading at each of the walls is equal to the tributary width between adjacent walls. Based on this, the shear at each wall is

Wall A: $V_A = 372^{\#/ft}(40') = 14,880^{\#}$

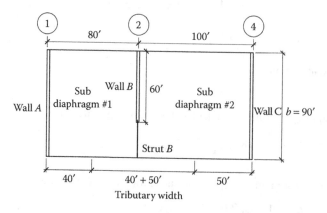

Figure 3.17 A 60 ft long shear wall is added to the building along gridline ②.

Wall B: $V_B = 372^{\#/ft}(40' + 50') = 33,480^{\#}$

Wall C: $V_C = 372^{\#/ft}(50') = 18,600^{\#}$

2. The unit shear across the width of the building along gridline ② is

$$V = \frac{V_B}{b} = \frac{33,480^{\#}}{90'} = 372^{\#/ft}$$

Then the force that the drag strut must collect and transfer to the shear wall on gridline ② is

$$F = 372^{\#/ft}(30') = 11,160^{\#}$$

3. The horizontal shear force at the top of the wall at gridline ④, taking into account the 1.6 load factor is

$$V_h = \frac{1.6(18,600^{\#})}{1000^{\#/k}} = 29.76^k$$

The unit shear along the length of the wall is

$$V_h = \frac{29.46^k}{90'} = 0.327^{k/ft}$$

Then, the shear force at each wall panel is

$$V_{h_{panel}} = 0.327^{k/ft}(8' \text{ wide panel})$$

$$= 2.62^k$$

The free body diagram of a wall panel is shown in Figure 3.18. Neglecting the weight of the panel and summary moments about 0 yields:

$$+\circlearrowleft \Sigma M_O = 0$$

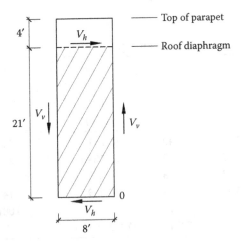

Figure 3.18 Free-body diagram of a wall panel.

$$2.62^k(21') - V_v(8') = 0$$

$$V_v = 6.87^k$$

Then the number of connections required between two panels is

$$\text{Number of connections} = \frac{V_v}{\phi V_h} = \frac{6.87^k}{4.8^k} = 1.43$$

Therefore, use two connections between panels.

3.4 SEISMIC STATIC FORCE PROCEDURE

3.4.1 Equivalent lateral force method

Example 3.5: Vertical distribution of seismic forces

For the five-story steel special moment frame (SSMF) building discussed in Example 2.4, determine the vertical distribution of seismic lateral forces (Figure 3.19 and Table 3.6).

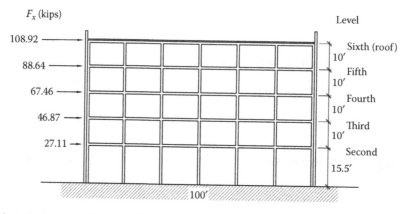

Figure 3.19 Seismic forces applied to the building.

Table 3.6 Vertical seismic force distribution of the base shear

				$K = 1.10$, $V = 331$ kips		
Level x	hx (ft)	hx (ft)	Wx (kips)	$w_x h_x k$ (kips-ft)	$C_{vx} = \dfrac{w_x h_x^k}{\sum_{i=1}^{n} w_i h_i^k}$	$F_x = C_{vx}V$ (kips)
6(roof)	55.5'	82.93	840	69,661.2	0.3049	100.92
5	45.5'	66.65	918	61,184.7	0.2678	88.64
4	35.5'	50.73	918	46,570.14	0.2038	67.46
3	25.5'	35.25	918	32,359.5	0.1416	46.87
2	15.5'	20.39	918	18,718.0	0.0819	27.11
I(main)			$\Sigma 4512$	$\Sigma 228,493.54$	1.000	331.0

The following information is given:

- Seismic weight of the building: $w = 4512$ kips
- Seismic response coefficient: $C_S = 0.0733$
- Seismic-force-resisting system

From ASCE-7 T 12.2-1:
C. Moment resisting frame systems:

1.   Steel special moment frames (SSMF)
Response modification coefficient $R = 8$
Over strength function $\Omega_0 = 3$
- Earthquake importance function: $I_e = 1.5$
- Design spectral acceleration: $S_{DS} = 0.47g$
- The fundamental period: $T = 0.691$ sec

As previously calculated in Example 2.4, the total design lateral force at the base of building is determined as follows:

$$V = C_S w = 0.0733(4512) = 331 \text{ kips}$$

Find F_x at each level
   The vertical distribution of F_x is determined according to

$$F_x = C_{vx}V \qquad\qquad (3.22) \text{ (ASCE 7 Equation 12.8-11)}$$

and

$$C_{vx} = \frac{w_x h_x^{k}}{\sum\limits_{i=1}^{n} w_i h_i^{k}} \qquad\qquad (3.23) \text{ (ASCE 7 Equation 12.8-12)}$$

where:
   C_{vx} is vertical distribution factor
   w_i and w_x is seismic weight at level i and x
   h_i and h_x is height from base to level i and x
   k is distribution exponent

when $T \leq 0.5$ sec, $k = 1.0$
when $T \geq 2.5$ sec, $k = 2.0$
   Else, when $0.5 < T < 2.5$, k can be conservatively taken as 2 or determined by linear interpolation between 1 and 2
   Interpolation for k:
   $0.5 < T = 0.691 < 2.5$

T	K
0.5	1.0
0.691	?
2.5	2.0

$$k_{T=0.691} = 1.0 + \frac{(0.691-0.5)}{2.5-0.5}\left[2.1 - 1.0\right] = 1.10$$

Example 3.6: Determine diaphragm design force

Determine the diaphragm design force F_{px} at each floor of the building. Floor and roof diaphragm seismic design forces are determined in accordance with

$$F_{px} = \frac{\displaystyle\sum_{i=x}^{n} F_i}{\displaystyle\sum_{i=x}^{n} w_i} w_{px} \qquad\qquad (3.24)\ (\text{ASCE 7 Equation 12.10-1})$$

where:
    F_{px} is the diaphragm design force
    F_i is the design force
    w_i is the seismic weight
    w_{px} is the seismic weight of diaphragm

The value of F_{px} should not be less than

$$F_{px} = 0.2 S_{DS} I_e w_{px} \qquad\qquad (3.25)\ (\text{ASCE 7 Equation 12.10-2})$$

Nor greater than

$$F_{px} = 0.4 S_{DS} I_e w_{px} \qquad\qquad (3.26)\ (\text{ASCE 7 Equation 12.10-3})$$

Check minimum $F_{px} = 0.2 S_{DS} I_e w_{px}$ (kips)

sixth level: $F_{p6min} = (0.2)(0.47)(1.5)(840) = 118.44$
    $118.44 > 100.92$
    Therefore 118.44kips controls

second–fifth level: $F_{p2-5} = (0.2)(0.47)(1.5)(918) = 129.4$
    $129.4 >$ all floors
    Therefore 129.4kips controls

Example 3.7: Determination of a diaphragm design seismic force low-rise building

For the building discussed in Example 3.4 the following seismic data apply:

- Seismic design category D
- Design spectral acceleration $S_{DS} = 1.0$
- Risk category II (ASCE 7 T 1.5-1)
  - Importance factor $I_e = 1.0$ (ASCE 7 T 1.5-2)
- Seismic-force-resisting system (ASCE 7 T 12.2-1)
  - Building frame system B
  - Intermediate precast shear walls response modification coefficient $R = 5.0$
- Redundancy factor $\rho = 1.0$
- The weight of the wood diaphragm construction $= 20^{\#/\text{ft}^2}$
- The weight of the precast panel $= 100^{\#/\text{ft}^2}$

The seismic weight w_{px} includes the weight of the diaphragm as well as the weight of the walls tributary to the diaphragm that are normal to the direction of the seismic force applied to the building (see Figure 3.20).

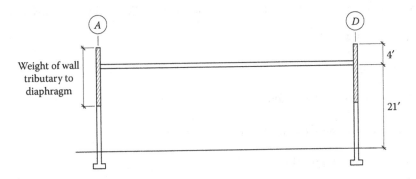

Figure 3.20 Section through the building.

The walls that are parallel to the applied seismic force are not typically considered in determining the tributary diaphragm weight, as the diaphragm does not support these walls.

The weight of the wall, which is tributary to the diaphragm (Figure 3.21) is calculated as follows based on a 1 ft width of wall.

Summing moments about the base of the wall to determine the seismic weight of the wall tributary to the diaphragm.

$$+\circlearrowright \; \Sigma M_A = 0$$

$$100(25)\frac{25}{2} - B(21) = 0$$

$$B = 1,488.1^{\#/ft} \text{ of wall}$$

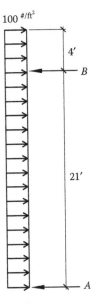

Figure 3.21 Free-body diagram of the mass of the wall.

The weight of the two walls normal to a seismic force, applied in the E–W direction, is

$$W_{wall} = 2 \times 1,488.1^{\#/ft} = 2976.2^{\#/ft}$$

The diaphragm is 90' wide,

$$W_{diaphragm} = 90' (20^{\#/ft^2}) = 1800^{\#/ft}$$

Then, the total weight of the diaphragm walls, w_{px}, is

$$W_{px} = (2976.2 + 1800)180 = 859.714^{\#}$$

$$W_{px} = \frac{859.714^{\#}}{1000^{\#/k}} = 859.7^{k}$$

The diaphragm force equation for a single story building is developed as follows:

$$F_{px} = \frac{\sum\limits_{i=x}^{n} F_i}{\sum\limits_{i=x}^{n} w_i} w_{px}$$

when $x = 1$, $\sum\limits_{i=x}^{n} w_i = w$

$$F_{px} = \frac{F_i}{w} w_{pi} \qquad (3.27)$$

From $F_x = C_{vx}V$

$$C_{vx} = \frac{w_x h_x^{\ k}}{\sum\limits_{i=x}^{n} w_i h_x^{\ k}}$$

For short periods $T < 0.5$ sec, $k = 1.0$
   Then $C_v = 1.0$
   And $F_1 = C_v V = F_1 = V$ \qquad (3.28)

$$V = e_s w = \frac{S_{DS} I_e}{R} w \qquad (3.29)$$

Then combining (3.27) and (3.28)

$$F_{p_1} = \frac{S_{DS} I_e}{R} w_{p_1}$$

Then

$$F_{p_1} = \frac{(1.0)(1.0)}{5.0} 859.7 = 172^{k}$$

Check limits:
   F_{p_1} *minimum*
   $0.2 S_{DS} I_e w_{px} = (0.2)(1.0)(1.0)859.7 = 172^{k}$
   Therefore $F_{p_1} = 172^{k}$ OK

3.5 HORIZONTAL AND VERTICAL IRREGULARITIES

Buildings are classified as regular or irregular as it pertains to their structural composition, that is, buildings that have horizontal and/or vertical irregularities, as defined in ASCE 7, Section 12.3, in association with their seismic design category are subject to specific design procedures and analyses.

3.5.1 Horizontal irregularities

There are five types of horizontal irregularities:

1a. *Torsional irregularity—to be determined when diaphragms are not flexible:* Torsional irregularity type 1a exists when the maximum story drift (including accidental torsion) at one end of the building exceeds 1.2 times the average of the story drift at both ends of the building

1b. *Extreme torsional irregularity—to be determined when diaphragms are not flexible:* Torsional irregularity type 1b exists when the maximum story drift (including accidental torsion) at one end of the building exceeds 1.4 times the average of the story drift at both ends of the building

2. *Reentrant corner irregularity:* A reentrant corner irregularity exists when the plan structural configuration of the lateral force resisting system contain reentrant corners where both plan projections of the reentrant corners exceed 15% of the total plan dimension of the building in the given direction

3. *Diaphragm discontinuity irregularity:* A diaphragm discontinuity irregularity exists when diaphragms have abrupt stiffness discontinuities either arising from open areas of portions of the gross area of the diaphragm comprising 50% or more or changes in stiffness of the diaphragm of more than 50% from one level to the next

4. *Out-of-plane offsets irregularity:* An out-of-plane offset irregularity exists when the lateral force resisting system's vertical elements have out-of-plane offsets

5. *Nonparallel system irregularities:* A nonparallel system irregularity exists when the vertical lateral force resisting system is not parallel to or symmetric about the major orthogonal axes

> Example 3.8: Horizontal irregularity type 1a and type 1b: torsional
> and extreme torsional irregularities
>
> The following displacements are given for a two-story rigid diaphragm SSMF building shown in Figure 3.22.
> Horizontal irregularities—Type 1a and 1b only apply to buildings with rigid and semi-rigid diaphragms.
> Determine if Type 1a or Type 1b torsional irregularity exists at the roof level.
> The irregularity check is defined in terms of story drift.

$$\Delta_x = (\delta_x - \delta_{x-1}) \tag{3.30}$$

A torsional irregularity exists at level x when the maximum story drift is greater than 1.2 times the average story drift:

$$\Delta_{max} > 1.2\Delta_{Avg.} = \frac{1.2(\Delta_{L,x} + \Delta_{R,x})}{2} \tag{3.31}$$

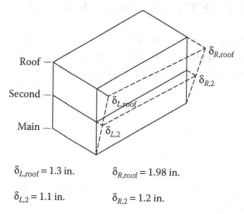

$\delta_{L,roof} = 1.3$ in. $\delta_{R,roof} = 1.98$ in.

$\delta_{L,2} = 1.1$ in. $\delta_{R,2} = 1.2$ in.

Figure 3.22 Lateral displacements of a two-story building.

Determining story drift at roof

$$\Delta_{L,roof} = (\delta_{L,roof} - \delta_{L,2}) = (1.3 - 1.1) = 0.2 \text{ in.}$$

$$\Delta_{R,roof} = (\delta_{R,roof} - \delta_{R,2}) = (1.98 - 1.2) = 0.78 \text{ in.}$$

$$\Delta_{avg.} = \frac{(0.2 + 0.78)}{2} = 0.49 \text{ in.}$$

$$\Delta_{max} = 0.78 > 1.2(0.49)$$

Thus a torsional irregularity exists—Type 1a
    Check for extreme torsional irregularity—Type 1b

$$\Delta_{max} = 0.78 > 1.4(0.49)$$

Thus an extreme torsional irregularity exists—Type 1b
    When a torsional irregularity exists at a level, the accidental torsional moment M_{ta} must be increased by an amplification factor A_x. A specific amplification factor is required at each level of the building.
    The amplification factor is a function of displacements, not the story drifts.

$$A_x = \left[\frac{\delta_{max}}{1.2\delta_{avg}} \right]^2 \qquad\qquad \text{(3.32) (ASCE 7 Equation 12.8-14)}$$

Then the amplification factor for the roof is calculated as shown here:

$$\delta_{max} = 1.98 \text{ in.}$$

$$\delta_{avg} = \frac{1.98 + 1.3}{2} = 1.64$$

Then,

$$A_x = \left[\frac{1.98}{1.2(1.64)} \right]^2 = 1.012 > 1.0$$

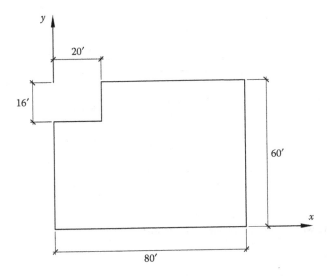

Figure 3.23 Plan of the building with a reentrant corner.

Example 3.9: Horizontal irregularity type 2: reentrant corner irregularity

The building shown in Figure 3.23 has a reentrant corner with plan projections in the x- and y-directions of 20 ft and 16 ft, respectively. Determine if a horizontal type 2 irregularity exists.

$$\frac{20'}{80'} = 0.25 > 0.15 \quad \text{of 80' plan dimension}$$

and,

$$\frac{16'}{60'} = 0.27 > 0.15 \quad \text{of 60' plan dimension}$$

Since both projections exceed 15% of their respective building dimension, then a type 2 reentrant corner irregularity exists.

Example 3.10: Horizontal irregularity type 3: diaphragm discontinuity irregularity

The floor plan of a multistory building is shown in Figure 3.24 with a cutout in the center of the floor.

The opening in the floor plate is

$$A_{open} = (125)(100) = 12,500 \text{ ft}^2$$

The percentage of the opening in relation to the gross floor plate is

$$\frac{12,500}{125 \times 175} = 0.57 > 0.50, \text{ or } 50\% \text{ of the total floor plate}$$

Hence, a diaphragm discontinuity irregularity exists.

It should be noted that because an irregularity may also exist if the stiffness of an adjacent floor has a change in stiffness of 50% or more, a comparison of the stiffness of the floor with the cutout, to the floor above or below in regard to stiffness should also be checked (Figure 3.25).

That is, if $\Delta_1 \geq 1.5\Delta_2$ then a diaphragm discontinuity exists.

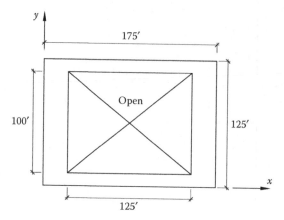

Figure 3.24 Floor plate with a cutout/opening in the center.

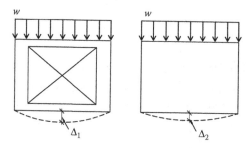

Figure 3.25 A comparison of deflections of two adjacent floors.

Example 3.11: Horizontal irregularity type 4: out-of-plane offsets irregularity

A three-story shear wall building shown in Figure 3.26 has shear walls, in the y- direction, on second and third floors along gridlines ① and ④. The shear wall, along gridline ④, continues to the first level, however on gridline ①, there are columns on the first level. The shear wall is located/shifted to gridline ②. This creates an offset of one bay in the vertical elements of the lateral force-resisting system.

Thus there exists an out-of-plane offset on gridline ① at the first level.

Example 3.12: Horizontal irregularity type 5: nonparallel system irregularity

The walls shown for the building in Figure 3.27 are part of the lateral resisting system. The lateral load resisting system located on gridline ⑤ is not parallel to or symmetric to the major orthogonal axes.

Thus a nonparallel irregularity exists.

Vertical irregularities
There are five types of vertical irregularities:

1a. *Stiffness—Soft Story Irregularity:* Soft story irregularity exists when the story stiffness is less than 70% of the story above or 80% of the average of the three stories above.

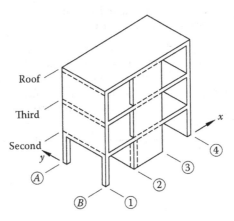

Figure 3.26 Three-story shear wall building.

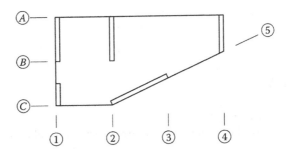

Figure 3.27 Plan view of a shear wall building.

1b. *Stiffness—Extreme Soft Story Irregularity:* Extreme soft story irregularity exists when the story stiffness is less than 60% of the story above or 70% of the average of the three stories above.

2. *Weight (mass) Irregularity:* Weight irregularity exists when the effective mass of any story is more than 150% of the effective mass of any adjacent story. Roofs, which are lighter than the floors below, are excluded.

3. *Vertical Geometric Irregularity:* Vertical geometric irregularity exists when the horizontal dimension of the lateral force resisting system in any story is more than 130% of that in an adjacent story.

4. *In-plane Discontinuity in Vertical Lateral-Force-Resisting Element Irregularity:* In-plane discontinuity in vertical lateral-force-resisting elements irregularity exists when there is an in-plane offset of a vertical seismic force resisting element greater than the length of the elements, which puts overturning demands on a beam, column, truss, or slab.

5a. *Discontinuity in Lateral Strength—Weak Story Irregularity:* Weak story irregularity exists when the story lateral strength is less than 80% of the strength above.

5b. *Discontinuity in Lateral Strength—Extreme Weak Story Irregularity:* Extreme weak story irregularity exists when the story lateral strength is less than 65% of the strength above.

Example 3.13: Vertical irregularity type 1a and 1b: soft story and extreme soft story irregularity

The intent of this irregularity check is to compare values of lateral stiffness of individual stories. It is common to compare displacements rather than the stiffness of the floors of the structure as the stiffness can be difficult to compute and the displacements can be readily obtained from computer generated analyses programs. When the story heights vary significantly from floor to floor, story drift ratios (the story drift divided by the story height) are a better comparison.

Since the stiffness is equal to the reciprocal of the displacement, $k = (1/\delta)$

Then the irregularity check

$$k_x < k_{x+1}$$

can be written as

$$\frac{1}{\delta} < 0.7 \frac{1}{\delta_{x+1}}$$

then rearranging the terms

$$\delta_{x+1} < 0.7(\delta_x)$$

or
when

$$70\% \text{ of } \delta_x > \delta_{x+1},$$

then a soft story irregularity exists.

Using story drift ratios the test for soft story irregularity is as follows:

1. When 70% of $\dfrac{\delta_x}{h_x}$ exceeds $\dfrac{\delta_{x+1} - \delta_x}{h_{x+1}}$

and

2. When 80% of $\dfrac{\delta_x}{h_x}$ exceeds $\dfrac{1}{3}\left[\dfrac{(\delta_{x+1} - \delta_x)}{h_{x+1}} + \dfrac{(\delta_{x+2} - \delta_{x+1})}{h_{x+2}} + \dfrac{(\delta_{x+3} - \delta_{x+2})}{h_{x+3}}\right]$

The story drift ratios of the building in Figure 3.28 are determined as follows:

$$\frac{\Delta_2}{h_2} = \frac{\delta_{2e} - 0}{h_2} = \frac{(0.75 - 0)}{144} = 0.0052$$

$$\frac{\Delta_3}{h_3} = \frac{\delta_{3e} - \delta_{2e}}{h_3} = \frac{(1.12 - 0.75)}{120} = 0.0031$$

$$\frac{\Delta_4}{h_4} = \frac{\delta_{4e} - \delta_{3e}}{h_4} = \frac{(1.57 - 1.12)}{120} = 0.0038$$

$$\frac{\Delta_5}{h_5} = \frac{\delta_{5e} - \delta_{4e}}{h_5} = \frac{(1.89 - 1.57)}{120} = 0.0027$$

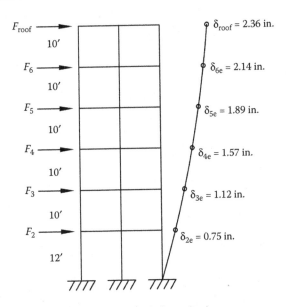

Figure 3.28 A steel special moment frame building with floor displacements.

$$\frac{\Delta_6}{h_6} = \frac{\delta_{6e} - \delta_{5e}}{h_6} = \frac{(2.14 - 1.89)}{120} = 0.0021$$

$$\frac{\Delta_{roof}}{h_7} = \frac{\delta_{re} - \delta_{6e}}{120} = \frac{(2.36 - 2.14)}{120} = 0.0018$$

Check the second floor (the first elevated floor):

$$0.70\left(\frac{\Delta_2}{h_2}\right) = 0.70(0.0052) = 0.0036 > \frac{\Delta_3}{h_3} = 0.0031$$

Thus a soft story irregularity exists.
   Check for an extreme soft story:

$$0.60\left(\frac{\Delta_2}{h_2}\right) = 0.60(0.0052) = 0.0031$$

Which is equal to the drift ratio of (Δ_3/h_3), thus OK:
   Check

$$0.70\left(\frac{\Delta_2}{h_2}\right) = 0.0036 > \frac{1}{3}[0.0031 + 0.0038 + 0.0027] = 0.0032$$

An extreme soft story irregularity exists.

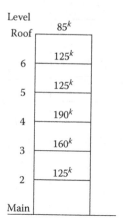

Figure 3.29 The seismic weights of a multilevel building.

Example 3.14: Vertical irregularity type 2: weight (mass) irregularity

A six-story building is shown in Figure 3.29. Determine if a type 2 weight irregularity exists.

At level 2:

$$1.5 \times w_2 = 1.5(125) = 187.5^k$$

At level 3:

$$1.5 \times w_3 = 1.5(160) = 240^k$$

At level 5:

$$1.5 \times w_5 = 1.5(125) = 187.5^k$$

then,

$$w_4 = 190^k > 1.5 w_5 = 187.5^k$$

Therefore, a weight irregularity exists.

Example 3.15: Vertical irregularity type 3: geometrical irregularity

For the four-story braced frame building shown in Figure 3.30. Determine if a type 3 irregularity exists.

The lateral force resisting system on the second level is two-bays wide and the lateral system on the third level is only one-bay wide. Then

$$\frac{\text{width of level 2}}{\text{width of level 3}} = \frac{50'}{25'} = 2.0$$

$$200\% > 130\%$$

Therefore a geometrical irregularity exists.

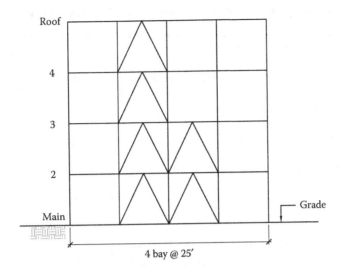

Figure 3.30 A four-story building with a braced frame lateral system.

Example 3.16: Vertical irregularity type 4: in-plane discontinuity in vertical lateral-force-resisting element irregularity

The offset of the vertical lateral force resisting system is 60 ft, which is greater than the bay width of the lateral system (Figure 3.31).

Thus, an in-plane discontinuity in vertical lateral force resisting element irregularity exists.

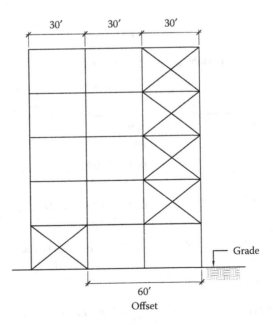

Figure 3.31 A multistory building with a braced frame lateral system.

Example 3.17: Vertical irregularity type 5a and 5b: weak and extreme weak story

The shear strengths are given for the piers of the three-story (concrete masonry unit [CMU]) masonry building shown in Figure 3.32, in Table 3.7. Determine if a weak story irregularity exists.

The total shear at each floor level is calculated here (Table 3.8):

$$V_{n_{3rd}} = 25 + 25 + 25 + 25 = 100^k$$

$$V_{n_{2nd}} = 19 + 19 + 19 + 19 = 76^k$$

$$V_{n_{main}} = 15 + 8 + 8 + 15 = 46^k$$

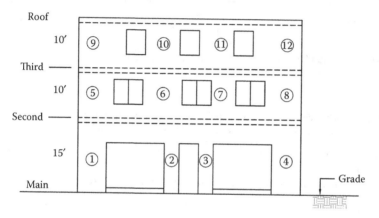

Figure 3.32 Labeled piers of a three-story masonry building.

Table 3.7 Calculation of diaphragm force at each level

		K = 1.10, V = 331 kips	
Level	w_i(kips)	F_i(kips)	F_{px}(kips)
6(roofs)	840	100.92	$\dfrac{100.92}{840}(840) = 100.92$
5	918	88.64	$\left(\dfrac{88.64 + 100.92}{840 + 918}\right)918 = 98.99$
4	918	67.46	$\left(\dfrac{67.46 + 88.64 + 100.92}{840 + 2(918)}\right)918 = 88.17$
3	918	46.87	$\left(\dfrac{46.87 + 67.46 + 88.64 + 100.92}{840 + 3(918)}\right)918 = 77.62$
2	918	27.11	
	$\Sigma 4512$	331	$\left(\dfrac{331}{4512}\right)918 = 67.34$

Table 3.8 Shear strength of piers as labeled in Figure 3.32

Pier #	Shear strength $V_n(k)$
1	15
2	8
3	8
4	15
5	19
6	19
7	19
8	19
9	25
10	25
11	25
12	25

Check story strength

$$V_{n2nd} < 0.80V_{n3rd}$$

$$46^k < 0.65(76^k) = 49.4^k$$

Thus, an extreme weak story irregularity exists.

The following Tables 3.9 and 3.10 summarizes the required analyses and design implications for buildings possessing irregularities and their assigned seismic design categories.

Table 3.9 Horizontal structural irregularities

Type of irregularity	Building's seismic design category	Summary of analysis and design implications
I(a) Torsion 1) Applies to rigid and semi rigid diaphragm buildings	D, E, F	1) Equivalent lateral force analysis isn't permitted. (T12.6-1) 2) Seismic design forces are to be increased by 25% for connections of diaphragms to vertical elements and collectors and for collectors and their connections to vertical elements. (12.3.3.4)
	B, C, D, E, F	1) Building must be analyzed using a 3-D model. (12.7.3) 2) Dynamic analysis may be required. (16.2.2)
	C, D, E, F	1) Torsional amplification factor Ax must be used at each level. (12.8.4.3) 2) The design story drift shall not exceed the allowable story drift in T 12.12-1. (12.12.1)
I(b) Extreme Torsion 1) Applies to rigid and semi rigid diaphragm buildings	E, F	Not permitted (12.3.3.1)
	D	1) Equivalent lateral force analysis is not permitted. (T12.6-1) 2) Seismic design forces are to be increased by 25% for connection of diaphragms to vertical elements and collectors and for collectors and their connections to vertical elements. (12.3.3.4)
	B, C, D	1) Building must be analyzed using a 3-D model. (12.7.3)
	C, D	1) Torsional amplification factor Ax must be used at each level. (12.8.4.3) 2) The design story drift shall not exceed the allowable story drift in T 12.12-1. (12.12.1)

(Continued)

Table 3.9 (Continued) Horizontal structural irregularities

Type of irregularity	Building's seismic design category	Summary of analysis and design implications
2. Reentrant corners	D, E, F	1) Seismic design forces are to be increased by 25% for connection of diaphragms to vertical elements and collectors and for collectors and their connections to vertical elements. (12.3.3.4)
3. Diaphragm discontinuity	D, E, F	1) Seismic design forces are to be increased by 25% for connection of diaphragms to vertical elements and collectors and for collectors and their connections to vertical elements. (12.3.3.4)
4. Out-of-plan offsets	B, C, D, E, F	1) Must be designed to resist the seismic load effects including over strength factor of 12.4.3. (12.3.3.3) 2) Building must be analyzed using a 3-D model. (12.7.3) 3) Dynamic analysis may be required. (16.2.2)
	D, E, F	1) Seismic design forces are to be increased by 25% for connection of diaphragms to vertical elements and collectors and for collectors and their connections to vertical elements. (12.3.3.4) 2) Equivalent lateral force analysis isn't permitted. (T12.6-1)
5. Nonparallel system	C, D, E, F	1) Orthogonal combination procedure is required. Design requires 100% of force for one direction plus 30% of forces in perpendicular direction. (12.5.3)
	B, C, D, E, F	1) Building must be analyzed using a 3-D model. (12.7.3) 2) Dynamic analysis may be required. (16.2.2)
	D, E, F	1) Equivalent lateral force analysis is not permitted. (T12.6-1)

Table 3.10 Vertical structural irregularities

Type of irregularity	Building's seismic design category	Summary of analysis and design implications
1(a) Soft story 1) Does not apply to one-story buildings 2) Does not apply to two-story buildings in B, C, D seismic categories	D, E, F	1) Equivalent lateral force analysis is not permitted. (T12.6-1)
1(b) Extreme soft story 1) Does not apply to one-story buildings 2) Does not apply to two-story buildings in B, C, D seismic categories	D, E, F E, F	1) Equivalent lateral force analysis is not permitted. (T12.6-1) 1) Not permitted. (12.3.3.1)
2. Weight(mass) Irregularity 1) Does not apply to one-story buildings 2) Does not apply to two-story buildings in B, C, D seismic categories	D, E, F	1) Equivalent lateral force analysis isn't permitted. (T12.6-1)

(Continued)

Table 3.10 (Continued) Vertical structural irregularities

Type of irregularity	Building's seismic design category	Summary of analysis and design implications
3. Vertical geometric	D, E, F	1) Equivalent lateral force analysis is not permitted. (T12.6-1)
4. In-plane discontinuity	B, C, D, E, F	1) Must be designed to resist the seismic load effects including over strength factor of 12.4.3. (12.3.3.3)
	D, E, F	1) Seismic design forces are to be increased by 25% for connections of diaphragms to vertical elements and collectors and for collectors and their connections to vertical elements. (12.3.3.4) 2) For structures with $T > 3.5T_s$, equivalent internal force analysis is not permitted. (T12.6-1)
5(a) Weak story	E, F	1) Not permitted. (12.3.3.1)
	D	1) Equivalent lateral force analysis is not permitted. (T12.6-1)
5(b) Extreme weak story	D, E, F	1) Not permitted. (12.3.3.1)
	B, C	1) Maximum permitted height is 30 ft and not over two stories unless designed using the over strength factor. (12.3.3.2)

Typically a building's irregularities are identified during design and analysis. The designer should carefully consider the implications when designing the lateral force resisting system to minimize as much as possible where appropriate. Often a reasonable modification can effectively eliminate an irregularity.

Chapter 4

Methods

4.1 FRAME ANALYSIS BY APPROXIMATE METHODS

Statically indeterminate structures are solved either by exact methods, utilizing elastic analyses, or by approximate methods, involving the use of simplifying assumptions. While there are many types of structural engineering software packages available, which are capable of solving very sophisticated structural systems, knowing some simple approximate methods to solve statically indeterminate structures is very helpful to the practicing engineer as a starting point for design or to check output from software packages. Approximate methods, which are done by hand, can often enlighten the designer's understanding of the structural stability and force balance of a proposed structure.

Solving determinant structures using principles of statics requires the system to have the same number of unknowns as equations of statics. That is, when there are more unknowns than equations, the structure is indeterminate. A system, which has three more unknowns than equations of statics, is said to be indeterminate to the third degree.

A building frame, which comprised two vertical and one horizontal member—a bent or portal, is statically indeterminate to the third degree. Since a single portal frame is indeterminate to the third degree, a rigid frame building, which is three stories tall and three bays wide and forms a nine portal building frame, has 27 degrees of indeterminacy, thereby making rigid frame buildings highly indeterminate. For this reason, their exact method solutions are time consuming due to the number of simultaneous equations needed to be solved, and are almost always analyzed by computer. Today, even simplified approximate methods are only used on very small systems and when appropriate due to the ubiquitous computer applications which can analyze and design these systems faster than one can perform an approximate method.

In order to solve indeterminate systems by an approximate method, assumptions are made to remove the number of unknowns or the degree of indeterminacy. There are many different methods available for making approximate analyses. The methods presented here are limited to gravity and lateral force analysis of frames.

4.1.1 Analysis of building frames for vertical loads

A useful approximate method for analyzing building frames for vertical loads is one in which assumptions are made that allow horizontal members to be analyzed as simple beams. Estimating the inflection point (IP) locations of the horizontal members of the frame to create simple beam spans between the IP's allows the moments and shear forces to be computed by simple statics. Then, the portion of the beam, remaining between the column and the IP of the beam, can be considered a cantilever as shown in Figures 4.1 and 4.2.

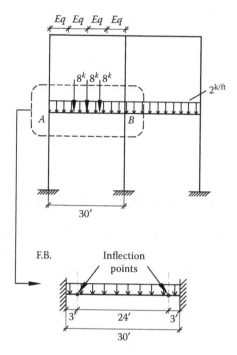

Figure 4.1 Assumptions associated with analyzing horizontal members for vertical loads in frames.

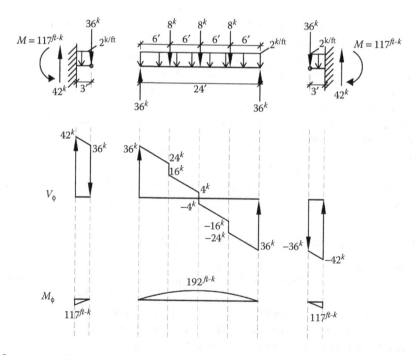

Figure 4.2 Separation of horizontal frame member into simple beam and cantilever.

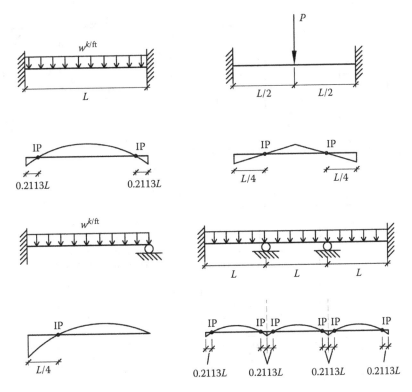

Figure 4.3 Inflection points of common beam configurations.

The IPs are taken at 1/10th of the span from the end of the beam. It is also assumed that there are no axial loads in the horizontal members. Tables of IP locations for common loading conditions and beam types, as those listed in Figure 4.3, are helpful for analysis.

4.1.2 Analysis of building frames for lateral loads

Probably, the most popular approximate method for calculating shear forces and moments in beams and columns on building frames due to lateral loads is the portal method. The portal method has been widely used by practitioners before the advent of computers due to its simplicity and accuracy and is still used today when appropriate.

In order to perform this method, there are several assumptions, which must be observed.

The first assumption is that the lateral forces applied to the frames are resisted by frame action—that is, the bending strength and fixity of the joints provide the resistance and stability of the frame. With this in mind, we accept that a laterally loaded frame will displace in such a manner that the moments in the columns and beams will be such as that shown in Figure 4.4.

The frame translation and joint rotation will generate influence points at mid-height of columns and mid-span of beams, and it is assumed that there is zero moment at the influence points in both the columns and beams. Therefore, the general assumption is that where there is zero moment in the structural system we can replace it with an internal hinge as shown for every column and beam in the two-story frame in Figure 4.5.

The second assumption is that interior columns attract twice the base shear force as compared to the exterior columns at each story of the frame, based on tributary area, also,

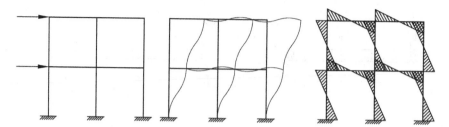

Figure 4.4 Assumed frame translation and moment diagram.

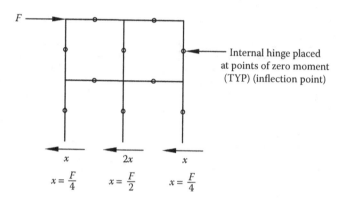

Figure 4.5 Internal hinge placements at mid-span of beams and mid-height of columns.

shown in Figure 4.5. The frame is now determinate and with the final assumption that only the exterior columns resist overturning of the frame, we can now solve with static equations of equilibrium.

The basic steps to perform a portal frame analysis are as listed here; however, each step has acceptable variation based on the designer's preference. In the examples that follow, certain deviations will be discussed.

Step 1: Calculate the column shears.
Calculate the base shears or horizontal reactions of the supporting columns based on the relationship of interior columns attracting twice the shear as the exterior columns on each of the levels.

Step 2: Make the structure determinant.
Place hinges at all the IPs of the columns and beams.

Step 3: Calculate the column moments.
The structure can now be separated at all of the hinges and each section of structure can be balanced with equations of statics to solve for the unknown forces. However, it should be noted for convenience, because the columns like the beams are assumed to have zero moment at their IPs located at mid-height of the columns and at mid-span of the beams, the columns' moments, top and bottom, equal the column's shear times half the column height.

Step 4: Calculate the beam moments and shears.
At any joint (not hinge) in the frame, the sum of the moments in the beams equals the sum of the moments in the columns. Since the column moments have been

previously determined, it is simple to calculate the beam moments by starting in the upper left and moving across to the right, adding or subtracting the moments accordingly. The points of inflection allow the beam shears to be calculated by dividing the beam moment by half the length of the beam.

Step 5: Axial forces in the columns and beams.

The axial forces in the columns are directly obtained from the beam shears. Similarly the axial forces in the beams are directly obtained from the column shears plus the lateral load.

Example 4.1: A one-story, two-bay lateral frame analysis using the portal method

Determine the reactions of the frame and the moments, shears, and axial forces for the beams and columns shown in Figure 4.6.

To begin an analysis of this two-bay, 6-degree indeterminate structure, simplifying assumptions must be applied to create a determinant system.

Step 1: Calculate the base shear of the columns where the interior columns attract twice that of the exterior columns (Figure 4.7).

Using equations of statics and summing the horizontal forces, we have

$$+ \rightarrow \Sigma F_y = 0$$

$$25 - x + 2x + x = 0$$

$$4x = 25$$

$$x = 6.25$$

Then, the base shear in the exterior columns is 6.25^k and the interior column base shear is $2 \times 6.25 = 12.5^k$.

Step 2: Place an internal hinge at IP locations at all beams and columns as shown in Figure 4.8.

Step 3: Calculate the column moments at the top and bottom of each of the columns. At the exterior columns (Figure 4.9):

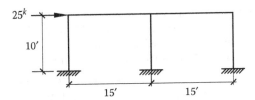

Figure 4.6 Two-bay moment frame with fixed support conditions.

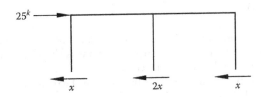

Figure 4.7 Base shear distributions at the columns of frame.

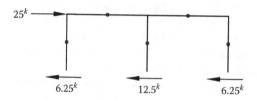

Figure 4.8 Frame with base shears and hinge placement.

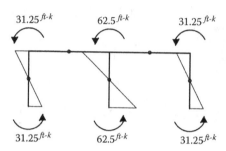

Figure 4.9 Calculated moments at the top and bottom of columns.

$$M = 6.25^k \left(\frac{10'}{2} \right) = 31.25^{k \cdot ft}$$

At the interior column:

$$M = 12.5^k \left(\frac{10'}{2} \right) = 62.5^{k \cdot ft}$$

Step 4: Calculate the beam moments by starting at the left side of the frame and balancing by summing the moments at each joint (Figure 4.10).
The joints A, B, and C are balanced as shown:

At joint A: $\Sigma M_A = 0$ At joint B: $\Sigma M_B = 0$ At joint C: $\Sigma M_C = 0$

$M_{Col} = M_{Beam\ AB}$ $M_{Col} - M_{Beam\ BA} = M_{Beam\ BC}$ $M_{Col} = M_{Beam\ CB}$

$31.25 = M_{Beam\ AB}$ $62.5 - 31.25 = M_{Beam\ BC}$ $31.25 = M_{Beam\ CB}$

                                $31.25 = M_{Beam\ BC}$

Now calculate the beam shears by dividing the beam moment by the half the length of the beam:

Shear in Beam AB: $V = \dfrac{M_{AB}}{l/2} = \dfrac{31.25}{15/2} = 4.17^k$

The shear is the same in both beams since we calculated the base shear in the columns. The shear for both columns and beams is shown in Figure 4.11.
Step 5: Calculate the axial forces in the columns and beams.
The axial forces are calculated directly from the shear in the beams and columns as shown in Figure 4.12a.
The reactions of the system are calculated and shown in Figure 4.12b.

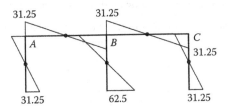

Figure 4.10 Balanced moments at joints A, B, and C.

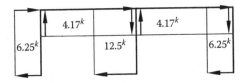

Figure 4.11 Beam and column shear diagram showing force direction.

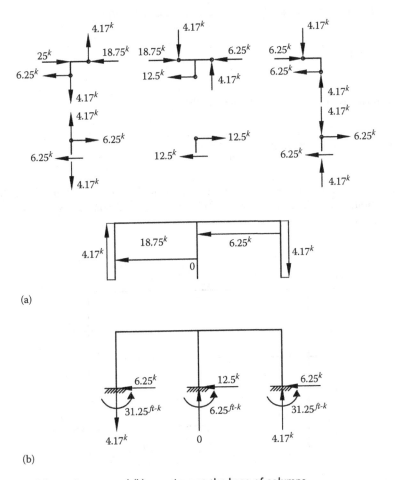

(a)

(b)

Figure 4.12 (a) Axial force diagram and (b) reactions at the base of columns.

Example 4.2: A two-story, four-bay frame analysis using the portal method

Determine the reactions of the frame and the moments, shears, and axial forces for the beams and columns shown in Figures 4.13 through 4.21.

Find the reactions:

$$x + 2x + x = (100^k + 100^k)$$

$$4x = 200^k$$

$$x = \frac{200}{4} = 50^k$$

$$V_{\text{Ext.}} = 50^k, \ V_{\text{Int.}} = 2(50^k) = 100^k$$

$$+\sum M_A = 0$$

$$+\sum M_A = 100(24') + 100(12') - C_y(40') - 300^{k \cdot ft} - 600^{k \cdot ft} - 300^{k \cdot ft}$$

$$C_y = \frac{3600 - 1200}{40} = 60^k$$

V at the second level $= x + 2x + x = 100^k$

V at the exterior column: $x = \frac{100}{4} = 25^k$

As you will notice there is no axial force in the middle column because the assumption, in this case, the overturning force is resolved completely by the exterior columns.

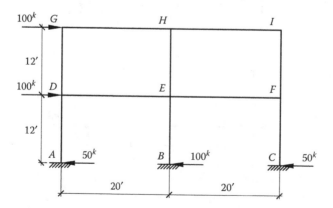

Figure 4.13 A two-story frame with lateral loads at the second and roof levels.

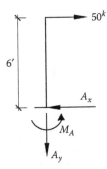

Figure 4.14 Free- body diagram at the column support, $M_A = 50^k(6') = 300^{k \cdot ft}$.

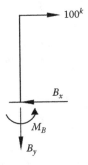

Figure 4.15 Free- body diagram at the column support, $M_B = 100^k(6') = 600^{k \cdot ft}$.

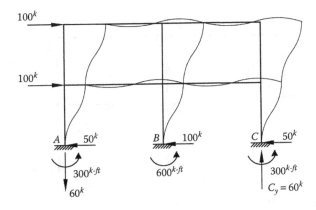

Figure 4.16 Deflected frame and reactions at supports.

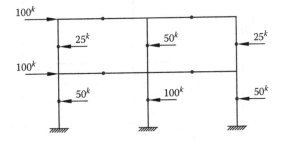

Figure 4.17 Shear in column at the main and second levels.

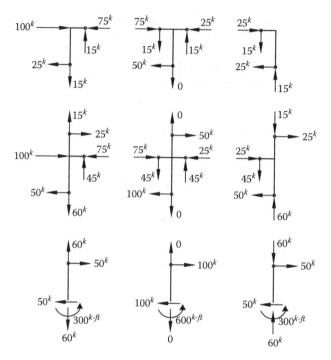

Figure 4.18 Diagram of all the parts of the frame separated at the hinges and the internal forces labeled.

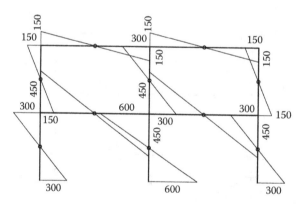

Figure 4.19 Moment diagrams of beams and columns.

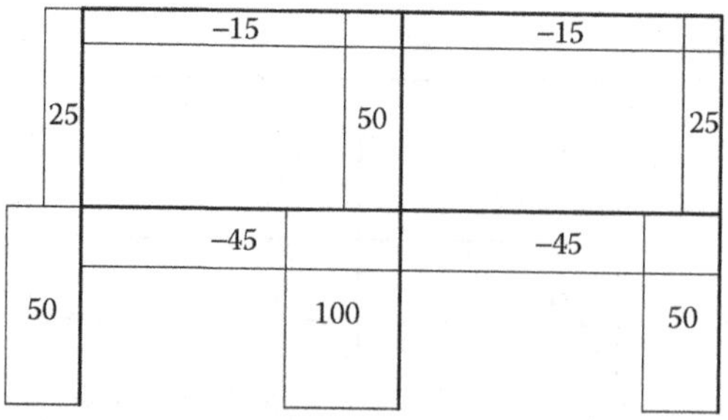

Figure 4.20 Shear diagrams of beams and columns.

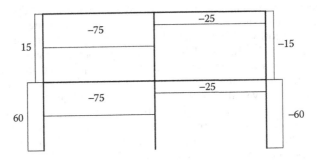

Figure 4.21 Axial force diagrams of beams and columns.

Example 4.3: Multistory frame with pin supports and varying heights, bay widths, and loads

Determine the reactions of the frame and the moments, shears, and axial forces for the beams and columns shown in Figure 4.22.

Step 1: Find the reactions of the system beginning with the shears at the base of the columns:

$$x + 2x + 2x + x = (20^k + 15^k + 10^k)$$

$$6x = 45^k$$

$$x = \frac{45}{6} = 7.5^k$$

$$V_{Ext.} = 7.5^k, \; V_{Int.} = 2(7.5^k) = 15^k$$

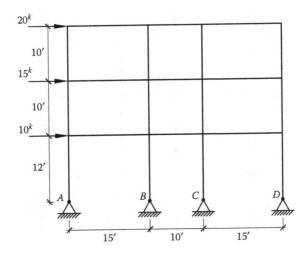

Figure 4.22 Frame with pin supports at the base.

In this example, there is no resistance to joint rotation at the base of the columns due to the pinned base supports, hence there are no moments generated at the base of the columns. It follows that the internal hinge which is placed at the mid-height of columns with fixed ends, is not done in this example for the first story columns because there does not exist an IP at mid-height in a pin-supported column (Figure 4.23).

The vertical reactions at the supports are best found once the beam shears are determined as we will show later in this example (Figure 4.24).

Separating the structure at its hinges and balancing the force using equations of statics accordingly can create the free-body diagram of the individual sections of the frame. Once the system's internal forces are balanced, the moments, shears, and axial forces can be determined (Figures 4.25 through 4.27).

First level: $7.5^k \left(\dfrac{12}{2} \right) = 45^{k \cdot ft}, 15^k \left(\dfrac{12}{2} \right) = 90^{k \cdot ft}$

Second level: $5.8^k \left(\dfrac{10}{2} \right) = 29^{k \cdot ft}, 11.7^k \left(\dfrac{10}{2} \right) = 58.5^{k \cdot ft}$

Third level: $3.3^k \left(\dfrac{10}{2} \right) = 16.5^{k \cdot ft}, 6.7^k \left(\dfrac{10}{2} \right) = 33.5^{k \cdot ft}$

The beam shears at the end of the beam are determined by dividing the beam moments by half the beam length as shown here (Figure 4.28):

First level: $\dfrac{74}{15/2} = 9.9^k, \dfrac{74}{10/2} = 14.8^k$

Second level: $\dfrac{45.5}{15/2} = 6.1^k, \dfrac{45.5}{10/2} = 9.1^k$

Third level: $\dfrac{16.5}{15/2} = 2.2^k, \dfrac{16.5}{10/2} = 3.3^k$

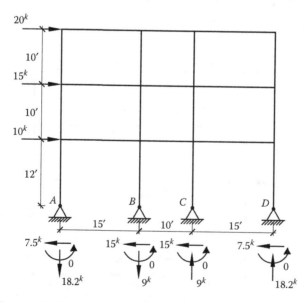

Figure 4.23 Moment frame with pin supports at the base.

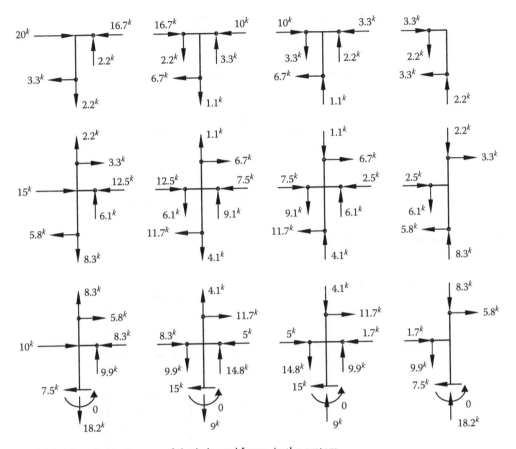

Figure 4.24 A free-body diagram of the balanced forces in the system.

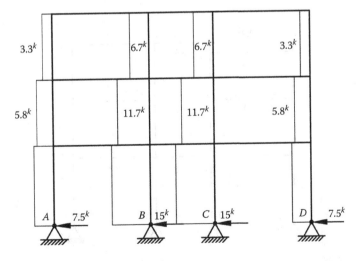

Figure 4.25 The column shear forces at each level.

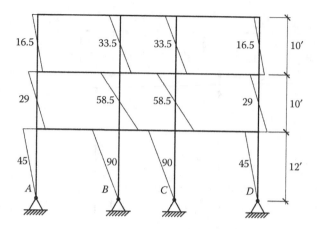

Figure 4.26 The column moments at each level.

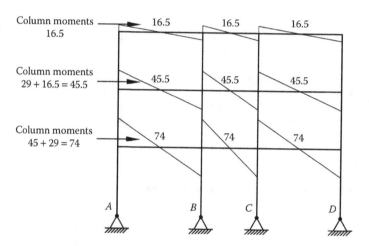

Figure 4.27 The beam moments balanced to the column moments at each joint.

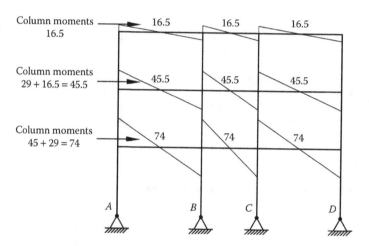

Figure 4.28 Beam shears diagram.

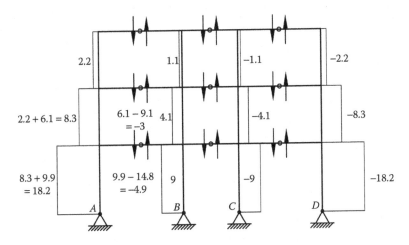

Figure 4.29 Column axial force diagram.

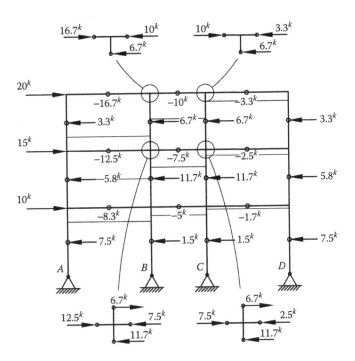

Figure 4.30 Axial forces in the beams.

The axial forces in the columns are determined from the beam shears connected to each column by collecting the loads from top to bottom at each column (Figures 4.29 and 4.30).

As shown with the other examples, there is a general process to the portal method procedure, along with its assumptions, and is able to be adapted and altered to solve varying systems. The designer should remain mindful to their objective, that is, the elements which they would like to solve and check their solution by going between a global and local system to confirm equilibrium. If used properly, the portal method can be a useful design aid.

Chapter 5

Designing and detailing of structures

5.1 LATERAL FORCE-RESISTING SYSTEMS

Permitted lateral seismic-force-resisting systems are described in ASCE 7, Section 12.2.1 and Table 12.2-1 *Design Coefficients and Factors for Seismic-Force-Resisting System*. The systems are divided into the following six basic groups:

1. Bearing wall systems
2. Building frame systems
   a. Utilizing braced frames
   b. Utilizing shear walls
3. Moment-resisting frame systems
4. Dual systems with special moment frames
5. Dual systems with intermediate moment frames
6. Cantilevered column systems

Each group is further divided by their specific vertical elements and type of materials, which resist lateral seismic loads. The structural systems in Table 12.2-1 also categorize according to detail requirements and Seismic Design Category regarding height limitations.

During the building design process, specifically the schematic design phase, various structural systems are often considered and vetted for a proposed building. The basics of a structural system is usually developed as part of the conceptual architectural design, that is, the architectural constraints that dictate the possible materials and spaces to be created also largely control the selection and design process of the structural systems for both gravity and lateral forces. Each of the six basic systems has certain advantages and disadvantages depending upon their application including the economics of each. Careful consideration is required when selecting a system for a proposed building.

5.1.1 Bearing wall systems

A bearing wall system supports both lateral and gravity loads, see Figure 5.1. A bearing wall works as a shear wall to resist lateral loads; in addition, it supports gravity loads. Bearing walls are of concrete, precast, or masonry construction or light-frame construction of wood or cold-formed steel. The gravity loads help to stabilize the walls against overturning. However, because bearing walls support both lateral and gravity loads, a failure of a wall will possibly create a failure in both lateral and gravity capacity (Figure 5.2).

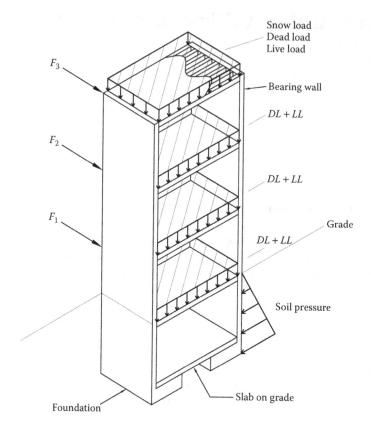

- Tilt-up construction
- Masonry
- Concrete
- Wood

Figure 5.1 Bearing wall systems.

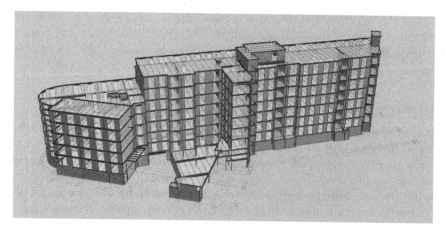

Figure 5.2 Bearing wall building modeled using structural analysis software.

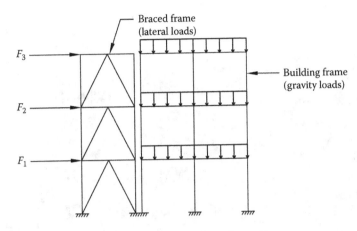

Figure 5.3 Building frame system utilizing a braced frame lateral system.

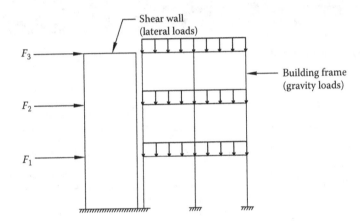

Figure 5.4 Building frame system utilizing a shear wall lateral system.

5.1.2 Building frame systems

A building frame system has separate or dedicated systems, which comprises a lateral-force-resisting element and a seismically separate gravity load supporting building frame. The lateral load-resisting portion of the system is either a braced frame or shear wall, see Figures 5.3 and 5.4. The building frame system is required to account for the maximum anticipated relative displacements between the two systems as per ASCE 7, Section 12.12.4. Braced frames are steel or steel and concrete composite construction. The shear walls are concrete, precast, composite steel and concrete, masonry, or light-framed construction. The building frame system is either steel or concrete (Figure 5.5).

5.1.3 Moment-resisting frame systems

Moment-resisting frames or rigid frames are specifically designed to provide support for both lateral and gravity loads, see Figure 5.6. Moment frames are much more flexible than braced frames, shear walls, and bearing walls and consequently displace and create large

Figure 5.5 Masonry bearing wall and building frame system.

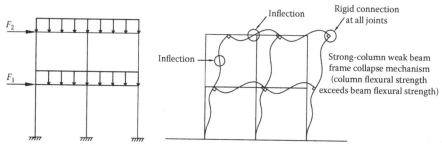

Figure 5.6 Moment-resisting frame systems.

drifts and p-delta effects which limit their performance. Detailing nonstructural elements such as attachments and cladding are crucial when accounting for deformations in moment frames (Figure 5.7).

5.1.4 Dual systems with special moment frames

A dual system provides a primary nonbearing lateral load-resisting system in the form of a braced frame or a shear wall coupled with a special moment-resisting frame system, which supports the gravity loads and at least 25% of the base shear, see Figure 5.8. Together, the two systems are required to resist the total base shear in proportion to their relative rigidities. Dual systems perform extremely well in high seismic regions and for this reason are permitted in all seismic design categories without limitation on height.

Figure 5.7 Moment frame building.

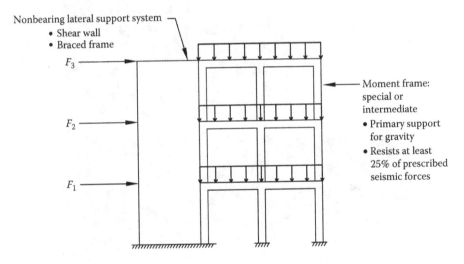

Figure 5.8 Dual systems.

5.1.5 Dual systems with intermediate moment frames

The intermediate moment frames such as the special moment frames must be designed to support at least 25% of the base shear; however, being intermediate frames, they cannot sustain the same seismic loading and as such are only permitted to be used in seismic design categories A, B, and C without limitations on height (Figures 5.9 and 5.10).

Figure 5.9 New steel and concrete building frame system being coupled with existing masonry bearing wall system.

Figure 5.10 Shear wall system.

5.1.6 Cantilevered column systems

A cantilevered column system is one in which the columns are restrained against rotation at the base usually with a bearing wall system above as shown in Figure 5.11. This system is also called an inverted pendulum structure and is defined as a structure, which has 50% of its mass at the top of a slender cantilever. This type of system has a low redundancy

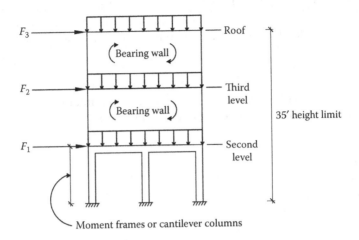

Figure 5.11 Cantilevered column systems.

Figure 5.12 Segmented masonry shear walls with building irregularity.

and overstrength. Like the bearing wall system, a failure in a lateral load-resisting element could cause a failure in the gravity system (Figures 5.11 and 5.12).

5.2 LOAD COMBINATIONS

The combining of loads in load combinations is governed by the International Building Code (IBC) Chapter 16 and in accordance with ASCE 7 where applicable. Those combinations presented in ASCE 7 are to be used when there is no applicable building code.

Load combinations are presented for both Load Resistance Factor Design (LRFD) or strength design and Allowable Stress Design (ASD) also known as working stress design.

The abbreviations for loads used in load combinations are as noted here:

D = dead load
E = earthquake load
F = load due to fluids with well-defined pressures and maximum heights
H = load due to lateral earth pressure, ground water pressure, or pressure of bulk materials
L = live load
L_r = roof live load
R = rain load
S = snow load
T = self-straining force
W = wind load

5.2.1 Load combinations using strength design or load resistance factor design

Where strength design or LRFD is used, buildings are required to be designed to resist the most critical loading condition produced by the following combination of factored loads according to IBC 1605.2:

1. $1.4(D + F)$
2. $1.2(D + F) + 1.6(L + H) + 0.5(L_r$ or S or R)
3. $1.2(D + F) + 1.6(L_r$ or S or R) $+ 1.6H + (f_1L$ or $0.5W)$
4. $1.2(D + F) + 1.0W + f_1L + 1.6H + 0.5(L_r$ or S or R)
5. $1.2(D + F) + 1.0E + f_1L + 1.6H + f_2S$
6. $0.9D + 1.0W + 1.6H$
7. $0.9(D + F) + 1.0E + 1.6H$

where:
$f_1 = 1$ for public assembly live loads in excess of 100 psf and parking garages and 0.5 for other live loads.
$f_2 = 0.7$ for roofs which hold snow and 0.2 for other roof configurations.

5.2.2 Load combinations using allowable stress design (basic load combinations)

Where working stress design or ASD is used, buildings are required to be designed to resist the most critical loading condition produced by the following combination of loads according to IBC 1605.3.

Loads are combined using the following nine equations:

1. $D + F$
2. $D + H + F + L$
3. $D + H + F + (L_r$ or S or R)
4. $D + H + F + 0.75(L) + 0.75(L_r$ or S or R)
5. $D + H + F + (0.6W$ or $0.7E)$

6. $D + H + F + 0.75(0.6W) + 0.75L + 0.75(L_r$ or S or R)
7. $D + H + F + 0.75(0.7E) + 0.75L + 0.75$ S
8. $0.6D + 0.6W + H$
9. $0.6(D + F) + 0.7E + H$

Alternative basic load combinations IBC 1605.3.2.

In lieu of the basic load combinations listed above, the IBC permits buildings to be designed to the most critical effects due to the following load combinations. When using these load combination that include wind or seismic loads, allowable stresses are permitted to be increased or load combinations be reduced where permitted by the material chapter of the IBC or the referenced standards.

Where wind loads are calculated in accordance with ASCE 7 Chapter 26 through 31, the coefficient (ω) is to be taken as 1.3. When using these alternative load combinations for foundation design, the loadings, which include vertical seismic load effect E_v, in equation 12.4-4 of ASCE 7 is permitted to be taken equal to zero.

1. $D + L + (L_r$ or S or R)
2. $D + L + 0.6\omega W$
3. $D + L + 0.6\omega W + S/2$
4. $D + L + S + 0.6\omega W/2$
5. $D + L + S + E/1.4$
6. $0.9D + E/1.4$

These load combination methods are discussed in further detail in the IBC and as required in ASCE 7. The reader is encouraged to reference the appropriate building code and design standard while using this book.

Example 5.1: Application of load combinations

For the steel special moment frame building shown in Figure 5.1, determine the maximum and minimum column loads on the main level (Figure 5.13).

Service loads:	Seismic lateral loads shown in figure
$w_{DR} = 1.2^{k/ft}$	
$w_{LR} = 0.6^{k/ft}$	$\rho = 1.0$
$w_D = 1.2^{k/ft}$	$S_{DS} = 1.0$
$w_L = 1.5^{k/ft}$	$F_1 = 0.5$

Factored load combinations as per ASCE 7, Section 2.3.2:

1. $1.4D$
2. $1.2D + 1.6L + 0.5(L_r$ or S or R)
3. $1.2D + 1.6(L_r$ or S or R) + F_1L or $0.5W$
4. $1.2D + 1.0W + F_1L + 0.5(L_r$ or S or R)
5. $1.2D + 1.0E + F_1L + 0.2S$
6. $0.9D + 1.0W$
7. $0.9D + 1.0E$

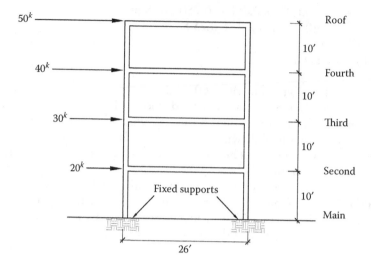

Figure 5.13 Four-story building with vertical distribution of lateral seismic loads.

where:
  D is the dead load
  L is the live load
  L_r is the live load roof
  S is the snow load
  R is the rain load
  E is the earthquake load
  W is the wind load
  F_1 is the load factor on L
  $F_1 = 0.5$ when $L \leq 100$ psf, else
  $F_1 = 1.0$

The governing load cases for D, L, L_r, and E are 2, 5, and 7.
  Wind loads and seismic loads need not be considered to act simultaneously.

Solution
The gravity loads are collected as follows (Figures 5.14 and 5.15):

$$D = \frac{[w_{DR} + w_D(3 \text{ floors})](\text{width of building})}{2 \text{ columns}} = \frac{(1.2 + 1.2(3))26}{2} = 62.4^k$$

$$L_r = \frac{w_{LR} \times 26'}{2} = \frac{0.6(26)}{2} = 7.8^k$$

$$L = \frac{(w_L \times 26')3 \text{ floors}}{2} = \frac{1.5(26)(3)}{2} = 58.5^k.$$

The seismic loads at the main level columns are found using the portal method.
  Shear at base:

$$x + x = 20 + 30 + 40 + 50$$
$$2x = 140$$
$$x = 70^k$$

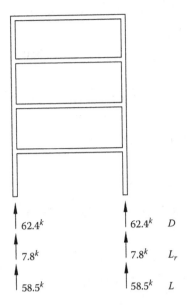

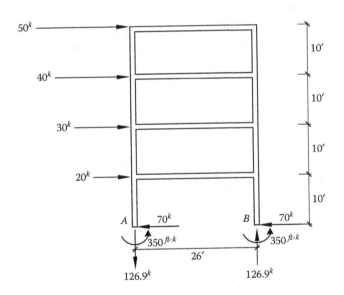

Figure 5.14 Reactions at base of footing due to gravity loads.

Figure 5.15 Reactions at base due to lateral seismic loads.

Therefore, $A_x = B_x = 70^k$.
   The moment at base:

$$M_A = M_B = 70^k \left(\frac{10'}{2} \right) = 350^{ft \cdot k}$$

Then, the seismic compression force in the column is found by summing moments about A:

$$+\curvearrowright \Sigma M_A = 0 = 20(10) + 30(20) + 40(30) + 50(40) - 350 - 350 - B_y(26)$$
$$B_y = 126.9^k$$

The seismic load combination from ASCE 7, Section 12.4.2.3 are used in lieu of those of Section 2.3.2

5. $(1.2 + 0.2S_{DS})D + \rho Q_E + F_1 L + 0.2S$
7. $(0.9 - 0.2S_{DS})D + \rho Q_E$

then,

5. $\left[1.2 + 0.2(1.0)\right]62.4 + 1.0(126.9) + 0.5(58.5) = 243.5^k$
6. $\left[0.9 - 0.2(1.0)\right]62.4 + 1.0(-126.9) = -83.22^k$
7. $(1.2)(62.4) + (1.6)(58.5) + 0.5(7.8) = 172.38^k$

Therefore, maximum $P_u = 243.5^k$ (compression)
and minimum $P_u = 83.22^k$ (tension).

5.3 BUILDING DRIFT

Example 5.2: Story drift determination

The elastic response deflection δ_{xe}, corresponding to the design seismic lateral forces, applied to the steel special moment frame (SSMF), are shown in Figure 5.16.

Risk Category II
$I_e = 1.0$
SSMF (T 12.2-1)
$R = 8, \Omega_0 = 3, C_d = 5.5$

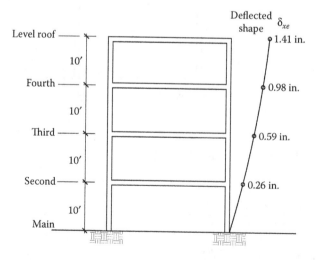

Figure 5.16 Four-story building showing elastic response deflections at each level.

Table 5.1 The elastic response deflection and
maximum inelastic story drift at each level

Level	δ_{xe} (in)	δ_x (in)
Roof	1.41	7.75
4th	0.98	5.39
3rd	0.59	3.25
2nd	0.26	1.43

Determine the maximum inelastic story drift of the roof level and check the story-drift limit (Table 5.1).

$$\delta_x = \frac{C_d \delta_{xe}}{I_e} \text{ (ASCE 7 12.8-15)}$$

$$= \frac{5.5 \delta_{xe}}{1.0} = 5.5 \delta_{xe}$$

Therefore,

$$\Delta = \delta_{x+1} - \delta_x$$
$$\Delta_{roof} = 7.75 - 5.39 = 2.37 \text{ in.}$$

From ASCE 7 T 12.12-1

Risk Category II
$\Delta_a = 0.025 h_{sx}$
When h_{sx} = story height

then, $\Delta_a = 0.025(10 \times 12) = 3.0$ in.
Therefore, $\Delta_{roof} = 2.37 < 3.0$ in OK.
The story drift is within the limit.

5.4 REDUNDANCY FACTORS

The redundancy factor ρ is either 1.0 or 1.3 for building structures. This value is based on the redundancy of the design of the lateral load-resisting system. That is, a failure in the lateral load-resisting system would be accounted for by a redundancy of the system's design. Failure to have sufficient redundancy will require the structural design to be penalized by having to use a value of 1.3. A redundancy factor must be assigned to each of the two orthogonal directions of the building.

Example 5.3: Determination of redundancy factor

Determine the redundancy factor in each direction for the system shown (Figure 5.17).

North–South direction:
   $\rho = 1.0$: because the system meets the requirement of having two bays of seismic resisting perimeter framing on each side of the building.

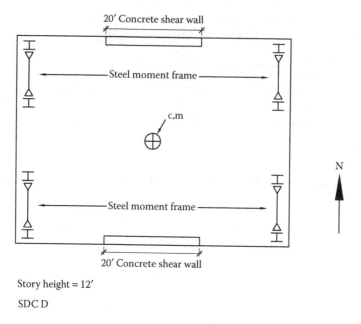

Story height = 12'

SDC D

Figure 5.17 Floor plan of building with North–South moment frames and East–West shear walls.

East–West direction:
The number of bays on the north and south sides of the building is calculated as:

$$\text{No. of bays} = \frac{\text{length of wall}}{\text{storyheight}} = \frac{20}{12} = 1.67 \text{ bays} < 2 \text{ bays}$$

Therefore, the east–west direction does not meet the requirement of two bays per side; hence, $\rho = 1.3$.

5.5 OVERSTRENGTH

The overstrength factor Ω_0 obtained in ASCE 7 T 12.12-1 for the different seismic lateral load-resisting systems approximates the inherent overstrength in the system. The special seismic loads, factored by the Ω_0 coefficient, are an approximation of the maximum force these elements are likely to experience.

5.6 STRUCTURAL SYSTEMS INTEGRATION

It is often necessary to design structural systems of buildings with different combinations of lateral-force-resisting systems. The combinations can be vertical as well as horizontal and are usually based on the demands of architectural constraints. That is, an architectural design for a proposed building may incorporate multiple uses and impose a different series of constraints such as those associated with a residential tower on top of an office and mixed commercial and retail space. The structural design for such a project may require the integration of several structural systems.

The following general rules regarding the selection and design of lateral-force-resisting systems are presented in Sections 12.2.2 and 12.2.3 of the ASCE 7 standard.

- A different seismic-force-resisting system is permitted in each of the two orthogonal directions. The limitations of each system apply for that system as well as the respective values for R, Ω_0, and C_d.
- When different seismic-force-resisting systems are used in combination and in the same direction, with the exception of dual systems, the most stringent values for R, Ω_0, and C_d are to be used for design.
- Where a structure has a vertical combination of systems in the same direction: (1) when the lower system has a lower Response Modification Coefficient, R, the design coefficients (R, Ω_0, and C_d) of the upper and lower systems are permitted to be used to calculate the forces and drifts of their respective systems; however, forces transmitted from the upper system to the lower system must be increased by multiplying by the ratio of the upper Response Modification Coefficient to the lower Response Modification Coefficient, and (2) when the upper system has a lower Response Modification Coefficient, the design coefficients (R, Ω_0, and C_d) for the upper system shall be used for both systems.

Example 5.4: Horizontal combination of systems in the same direction

Determine the appropriate response modification factors and design coefficients for the lateral structural system shown in Figure 5.18.

- The building is assumed to have rigid diaphragm floors.
- Special reinforced concrete building frame shear walls (SRCSWs) and steel special moment frames (SSMFs) provide lateral load resistance in the same direction.

From ASCE-7 T 12.2-1
SRCSW

$$R = 6, \ \Omega_0 = 2\frac{1}{2}, \ C_d = 5$$

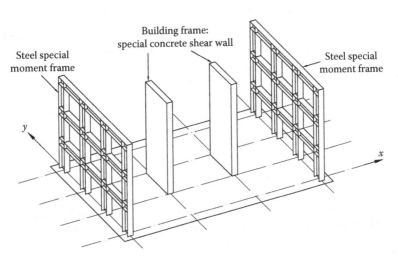

Figure 5.18 Three-story building with steel special moment frames and special concrete shear walls in the same direction.

Figure 5.19 Bearing wall and frame building with horizontal and vertical irregularities.

SSMF

$$R = 8, \Omega_0 = 3, C_d = 5\frac{1}{2}$$

The most stringent values for R, Ω_0, and C_d are to be used for both systems (Figure 5.19).

They are $R = 6, \Omega_0 = 3, C_d = 5\frac{1}{2}$.

If designed as a dual system, $R = 7, \Omega_0 = 2\frac{1}{2}, C_d = 5\frac{1}{2}$.

Example 5.5: Vertical combination of systems in the same direction

Determine the appropriate response modification factors and seismic design coefficients for the vertical combination of lateral structural system shown in Figure 5.20.

From ASCE-7 T 12.12-1
SSMF

$$R = 8, \Omega_0 = 3, C_d = 5\frac{1}{2}$$

SCBF

$$R = 6, \Omega_0 = 2, C_d = 5$$

When the upper system has a lower response modification coefficient, the design coefficients for the upper system should be used for both systems (Figure 5.21).

Hence,

$$R = 6, \Omega_0 = 2, C_d = 5$$

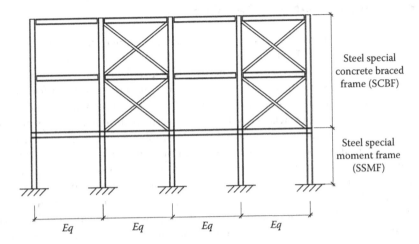

Figure 5.20 Vertical combination of seismic lateral load-resisting systems.

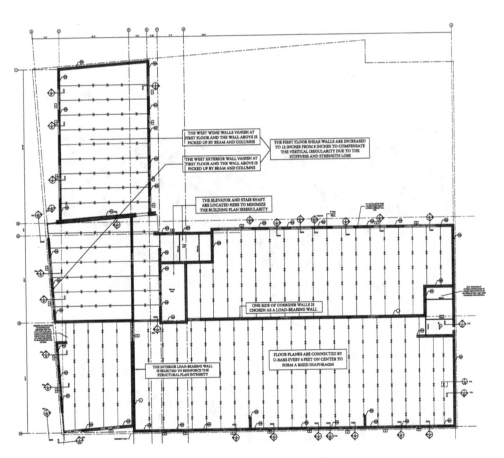

Figure 5.21 Concrete plank floor system with noted areas of irregularities.

5.7 SERVICEABILITY CONSIDERATIONS

To provide a properly functioning building, structural designers must keep serviceability goals in mind during design. That is, the building's performance as it pertains to building movements in the form of expansions and contractions, deflections, vibrations, and drifts are integral to a building's performance in regard to maintenance and comfort during normal usage. Serviceability issues, as we have discussed thus far in regard to structural irregularities, is not all inclusive. Serviceability from a structural designer's perspective extends to the long-term performance of the building. For example, the vertical deflections of floor systems should be designed such that in addition to meeting code requirement limits, they also function properly and do not diminish connection strength or force transfer and such issues. In addition, anticipated lateral movements can be well within design limits but still create adverse conditions if not properly detailed. A building undergoing lateral deformations, which creates wall deflections and diaphragm displacements, can possibly distort and adversely compromise the cladding and cause unwanted stress concentration, which can cause failure of the cladding system under normal use.

The designer must keep potential issues like this, as well as those that may arise from long-term irreversible deformations such as creep. All materials creep, and the potential for a deformed member to become unstable due to creep should be evaluated and perhaps be met with a redundant system to allow monitoring and periodic maintenance to be performed to combat any instability that may develop.

Other issues to be mindful of are presented here.

- Mechanical equipment, which can produce cyclic vibrations and cause fatigue to structural members, should be evaluated and dampened accordingly.
- Building structural systems should be evaluated and designed to minimize vibrations due to wind loading that may cause occupant discomfort. Very flexible buildings are

Figure 5.22 Single-story frame system integrated with wood framing.

susceptible to large movements and possible large vibrations, which should be damp-ened accordingly.

- Camber requirements, which are necessary for the structural design of members, also need to be coordinated with other trades so as to align properly when in place and be adequately integrated into the construction assembly.
- Dimensional changes to the structure due to thermal expansion and contraction can cause extensive damage to a building assembly. Careful analysis and study of potential movements of the structure should be performed and design considerations given to the location of expansion joints or the design of slotted connections which will allow for thermal growth.
- A building's durability should be designed such that environmental effects should be tolerated and endured to so as to maintain its function and composition under reason-able long-term maintenance (Figure 5.22).

Chapter 6

Steel

6.1 INTRODUCTION TO LATERAL STEEL DESIGN

The American Institute of Steel Construction (AISC) is a technical institute that supplies specifications, codes, research, quality certification, standardization, and technical assistance to its members and the general civil engineering community. The AISC Steel Construction Manual is the general guide for steel design and construction in the United States and it is the AISC Specifications for Structural Steel Buildings 360-10, which governs the design of steel members and connections.

There are two design methods presented in the AISC specifications. Steel design is to be performed in accordance with either one of the provisions for these methods. The methods are (1) load resistance factor design (LRFD) and (2) allowable stress design (ASD). The required strength of structural members and their connections may be determined by elastic, inelastic, or plastics analysis for load combinations associated with LRFD or by elastic analysis for load combinations associated with ASD. In both methods, the available strength must exceed the required strength.

In the ASD method, the allowable strength must equal or exceed the required strength:

$$R_a \leq \frac{R_n}{\Omega}$$

where:

R_a is the required strength determined by analysis for the ASD load combinations
R_n is the nominal strength determined according to the AISC specifications
Ω is the safety factor given by the AISC specifications for a specific limit state

In the LRFD method, the available strength is referred to as the design strength and the provisions for LRFD are structured, so that the design strength must equal or exceed the required strength.

$$R_u = \varphi R_n$$

where:

R_u is the required strength determined by analysis for the LRFD load combinations
R_n is the nominal strength determined according to the AISC specifications
φ is the resistance factor given by the AISC specifications for a specific limit state

Structural steel design according to the AISC specifications for either design method—LRFD or ASD—is based on limit states design principles, which define the boundaries of structural usefulness in regard to a structure's ultimate strength limit state and serviceability limit state. Strength limit state is the ability of the structure to carry imposed loads and the safety of the public, and serviceability limit states relate to the performance of the structure under normal service conditions. The design provisions for steel provide guidance for the design of buildings so that no applicable strength or serviceability limit state is exceeded.

The AISC manual references load combination used for steel design that are described in ASCE 7; however, the load combinations listed in Chapter 16 of the International Building Code (IBC) will govern in jurisdictions that have adopted the IBC.

A successful design will be based on the availability of structural steel shapes. The designer should remain aware of the ability to secure the shapes, which are detailed in the construction documents so as to not develop a design for which materials for a project will be difficult to procure. The availability of structural shapes can be determined by checking the AISC database at www.aisc.org/steelavailability in which a list of the producers of specific shapes is supplied.

This chapter as with the subsequent chapters related to design using concrete, wood, and masonry is not intended to be all inclusive in regard to the principles and practice of design using a specific material, but to provide a continuity of concepts previously discussed in this text, and give the reader an opportunity to review these concepts, and apply them in situations to give a holistic presentation of analysis and design of structural systems. The reader is assumed to have basic working knowledge of design and is encouraged to use the codes and design standards referenced with this text in conjunction with working the problems presented in this and the following Chapters 7, 8, and 9.

Lateral steel design of buildings is a relatively new concept. Early building engineering design incorporated loads only due to gravity. Overturning forces were considered to be incidental and checked as an afterthought. Today lateral steel design is very specific and is largely dependent on seismic requirements, specifically the seismic design category (SDC) and the steel seismic-force-resisting system (SFRS) selected. Proportioning and detailing a structure to sustain the ground motion or shaking induced from an earthquake is the practice of seismic design. The AISC *Seismic Provisions for Structural Steel Buildings* has been developed to provide a guide to designing structures, which can sustain a maximum considered earthquake for the site where the building is located, as defined by ASCE 7, with a low probability of collapse. That is, the seismic design of steel structures allows the structural members to enter into the inelastic region of the material (steel) due to large deformational demands, and to sustain *controlled* damage, and not to create a catastrophic failure. This requires careful proportioning of the structural system, so that inelastic behavior occurs in preselected members that have appropriate section properties to sustain large deformations without loss of strength of the connections.

The applicable building code and structural design standards, which are relevant for steel design of buildings, are listed as follows in the order of precedence:

- International Building Code (IBC 2015 Edition)
- Minimum Design Loads for Buildings and Other Structures (ASCE 7-10)
- Steel Construction Manual
  - Specifications for Structural Steel Buildings (AISC 360-10)
- Seismic Design Manual
  - Seismic Provisions for Structural Steel Buildings (AISC 341-10)
  - Prequalified Connections (AISC 358-10)

The maximum considered earthquake as defined in ASCE 7 is an earthquake with an annual occurrence that will provide a uniform collapse risk of 1% probability in 50 years. The design intent of ASCE 7 is to assure that structures assigned to risk categories I and II have not greater than a 10% chance of collapse should they experience a maximum considered earth quake.

The seismic-force-resisting systems consist of braced frames and moment frames, and those systems that require special detailing are classified into three levels of expected inelastic response categories, they are ordinary, intermediate, and special, based on their level of ductility. Designing to meet the seismic requirements of the AISC *Seismic Provisions for Steel Structures* is mandatory for structural systems referenced in ASCE 7 T 12.12-1. Steel structures assigned to SDCD are typically required to meet the requirements of the seismic provisions. Braced and moment frame systems that do not require special detailing are governed by the design requirements of the AISC *Specification* and they are discussed in Part 3 of the AISC *Seismic Design Manual*. A popular misconception is that when seismic detailing of the seismic-force-resisting system is not required, there are no other seismic requirements. This is not true; structures assigned to seismic design categories B through F are subject to many other seismic design considerations such as those in ASCE 7:

- Horizontal and vertical structural irregularities
- Seismic load effects and combinations
- Direction of loading
- Amplification of accidental torsional moment
- Collector elements
- Foundation design

The response modification coefficient, R, provided for the various structural systems noted in ASCE 7 T12.12-1, is applied in the base shear equation as $1/R$, which establishes the level of strength required to resist seismic forces elastically that is permitted for design. A higher value of R results in a more ductile system, able to resist higher seismic forces.

For structures assigned to SDC B or C in ASCE 7, the design engineer is able to choose a system that has a value of $R = 3$ and use the AISC *Specifications* for designing or selecting a system with a higher value of R, which must be detailed in accordance with the AISC *Seismic Provisions*. Systems designed with a value of $R = 3$ will have ductility associated with conventional steel framing not specifically detailed for high seismic resistance. Generally, buildings that are designed to meet the AISC *Specifications* with $R = 3$ are more economical than the same structure designed with the AISC *Seismic Provisions* using a higher value of R. The additional cost of the structure is attributed to the extra time to fabricate, erect, and inspect in order to achieve the high ductility requirements. This additional associated cost more than offset the extra steel tonnage in material for the $R = 3$ system.

The SFRS that requires seismic detailing according to AISC *Seismic Provisions* are as follows:

- Ordinary concentrically braced frames (OCBF)
- Special concentrically braced frames (SCBF)
- Eccentrically braced frames (EBF)
- Buckling-restrained braced frames (BRBF)

and

- Ordinary moment frames (OMF)
- Special moment frames (SMF) and intermediate moment frames (IMF)

6.2 SPECIAL CONCENTRICALLY BRACED FRAME SYSTEMS

Special concentrically braced frame systems resist lateral loads through the axial strength of the braces. They are generally configured to dissipate energy by tension yielding and/or compression buckling in the braces. The connection of the braces must be proportioned to remain elastic as they undergo deformation (Figure 6.1).

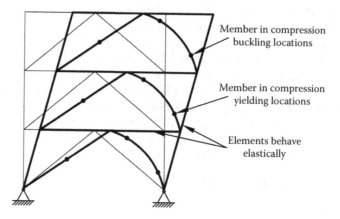

Member in compression
buckling locations

Member in compression
yielding locations

Elements behave
elastically

Figure 6.1 Locations or yielding and buckling of the SCBF system.

Example 6.1: Building design example of a special concentrically braced frame

This problem requires the determination of the seismic loads based on the location of the building and design of a steel special concentric-braced frame system based on AISC seismic provisions to resist seismic lateral loading. In order to put it all together, the following information must be obtained for analysis.

The office building shown in Figure 6.2 is in SDCD (Figure 6.3).

- $S_{DS} = 1.0$, $S_{D1} = 0.42$
- The floors are assumed to be rigid diaphragms
- The redundancy $\rho = 1.0$: two bays of seismic force resisting elements are provided on each side of the building and at the perimeter. Therefore, a $\rho = 1.0$ requirement is met
- For a SCBF system, the response modification factor/coefficient $R = 6$ and over-strength factor $\Omega_0 = 2.0$

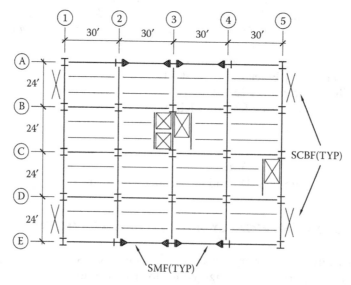

Figure 6.2 Plan view of a typical floor of a three-story office building with two bays of braced frames along grid lines 1 and 2, and two bays of moment frames along gridlines A and E.

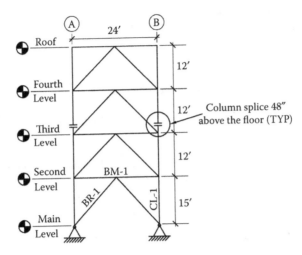

Figure 6.3 Elevation of typical braced frame along grid lines I and 5 shown in Figure 6.2.

Typical gravity loading is as follows:

- $D_{floor} = 85$ psf
- $D_{roof} = 70$ psf
- Curtain wall $= 200^{\#/ft}$ along the building perimeter at levels 2–4, and $120^{\#/ft}$ at the roof
- $L_{floor} = 80$ psf
- $S = 20$ psf
- $T_L = 8$ s

The following steps are required:

1. Determine the period of structure
2. Determine the seismic response coefficient
3. Determine the seismic weight of structure
4. Determine the seismic base shear
5. Vertical distribution of seismic forces to each level
6. Determine the seismic design story shear
7. Horizontal distribution to vertical elements of seismic-force-resisting system (rigid diaphragm analysis)
8. Design SCBF Brace BR-1
9. Design SCBF Column CL-1
10. Design SCBF Beam BM-1

Step 1: Determine the period of the structure T.

$$T_a = C_t h_n^x \hspace{3cm} (2.10) \text{ (ASCE 7 Equation 12.8–7)}$$

From table ASCE 7 T 12.8–2
Structure type: SCBF

$C_t = 0.02$

$x = 0.75$

The height of the building, $h_b = 51$ ft
Then the approximate fundamental period:

$$T_a = (0.02)(51)^{0.75} = 0.382 \text{ s}$$

Step 2: Determine the seismic response coefficient, C_s.

$$C_s = \frac{S_{DS}}{(R/I_e)} \qquad\qquad (2.12) \text{ (ASCE 7 Equation 12.8–8)}$$

An office building is assigned to risk category II, which corresponds to a seismic importance factor $I_e = 1.0$, from ASCE T 1.5-2.
Then,

$$C_s = \frac{S_{DS}}{T(R/I_e)} = \frac{1.0}{(6/1.0)} = 0.167$$

and C_s does not need to be greater than

$$C_s = \frac{S_{D1}}{T(R/I_e)} = \frac{0.42}{0.382(6/1.0)} = 0.183 \text{ for } T \le T_L$$

$$C_s = \frac{S_{D1}T_L}{T^2(R/I_e)} = \frac{0.42(8)}{(0.382)^2(6/1.0)} = 0.436 \text{ for } T > T_L$$

However, C_s should not be less than

$$C_s = 0.044S_{DS}I_e \qquad\qquad (2.13) \text{ (ASCE 7 Equation 12.8–5)}$$

$$= 0.044(1.0)(1.0) = 0.044$$

Therefore, $C_s = 0.167$
Step 3: Determine the seismic weight as tabulated in Table 6.1.
Step 4: Determine the seismic base shear.

$$V = C_sW$$

$$V = 0.167(4055) = 677^k$$

Step 5: Determine the vertical distribution of seismic force to each level.

$$F_x = C_{vx}v \qquad\qquad (3.23) \text{ (ASCE 7 Equation 12.8–11)}$$

Table 6.1 Tabulation of seismic weight for each floor of the building

Level	Area (Ft²)	DL (ksf)	Curtain wall weight (k)	Total W_i (k)
Roof	96 × 120 = 11.520	0.07	0.120 × 432′ = 51.84	858.24
4th	11,520	0.085	0.200 × 432 = 86.4	1065.6
3rd	11,520	0.85	86.4	1065.5
2nd	11,520	0.85	86.4	1065.6
				w = 4055.0

Table 6.2 Tabulation of seismic forces for each floor of the building

Level x	H_x (Ft)	W_x(kips)	$w_x h_x(k-ft)$	$C_{vx} = \dfrac{w_x h_x}{\sum_{i=1}^{n} w_i h_i}$	$F_x = C_{vx}V$(kips) $V = 677k$
Roof	51	858.24	43,770.24	0.3365	227.81
4th	39	1065.6	41,535.00	0.3194	216.23
3rd	27	1065.6	28,771.2	0.2212	149.75
2nd	15	1065.6	15,984.0	0.1229	83.2
			$\sum 130,060.44$	1.000	677.0

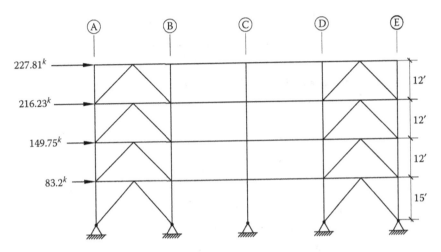

Figure 6.4 Seismic story forces applied to frame at each level.

$$C_{vx} = \frac{w_x h_x^{k}}{\sum_{i=1}^{n} w_i h_i^{k}}$$

(3.24) (ASCE 7 Equation 12.8–12)

For $T = 0.382s < 0.5s$, $k = 1.0$

The vertical distribution of seismic forces is tabulated in Table 6.2 (Figure 6.4).

Step 6: Determine the seismic design story shear, see Figure 6.5.

Step 7: Determine the horizontal distribution of seismic forces to the vertical elements of seismic-force-resisting system (rigid diaphragm analysis).

1. The center of mass and center of rigidity are coincidental for this building's structural systems' configuration.
2. For simplicity, ignore the effects from torsional forces due to accidental eccentricity.
3. Assume that seismic forces at each level are horizontally distributed evenly to the braced frames on grid lines ① and ⑤.

Then, the story force at each braced frame is equal to the story force divided by the number of frames (Figure 6.6; Table 6.3).

6.2.1 Brace design

The first level brace design is a good first step in the analysis and design procedure for a special concentric-braced frame system. The design forces can be determined from statics

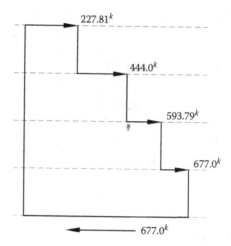

Figure 6.5 Seismic story shears at each floor level including roof.

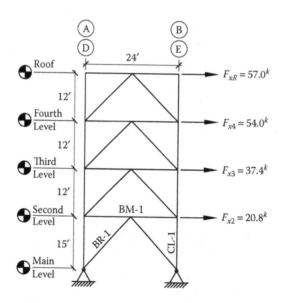

Figure 6.6 Horizontal seismic forces at each building level applied at each frame.

Table 6.3 Story forces at each braced frame at each level

Level	$F_x(k)$
Roof	$F_{xR} = 227.18/4 = 57.0$
Fourth	$F_{x4} = 216.23/4 = 54.0$
Third	$F_{x3} = 149.75/4 = 37.4$
Second	$F_{x2} = 83.2/4 = 20.8$

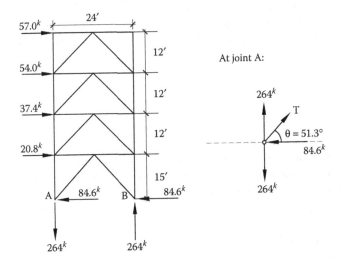

Figure 6.7 Reactions at base due to horizontal seismic forces.

or from more robust methods using computer-generated modeling methods. Typically, a 3D model representation of the entire building would be created to analyze the various load cases and from which the most severe effects would be evaluated for design of the braces.

Step 8: Now, we can begin the design of the SCBF brace BR-1
The seismic axial force in brace BR-1, shown in Figure 6.3, due to the horizontal seismic forces applied to the frame is determined by a static analysis (Figure 6.7).

$$V_{base} = 57.0 + 54.0 + 37.4 + 20.8 = 169.2^k$$

$$B_x = A_x = \frac{169.2^k}{2} = 84.6^k$$

$$+\Sigma M_A = 57(51) + 54.0(39) + 37.4(27) + 20.8(15) - B_y(24) = 0$$

$$B_y = \frac{6334.8}{24} = 264$$

At joint A:

$$+\Sigma F_y = 0$$

$$T \cos 51.3 - 84.6 = 0$$

$$T = 135.4^k$$

Thus, $Q_E = \pm 136^k$

From a separate gravity analysis, the axial loads in the brace due to dead and live loads are as follows:

$$P_{DL} = 14 \text{ kips}$$

$$P_{LL} = 13 \text{ kips}$$

The axial load due to snow load is negligible.

Select an ASTM A500 Grade B round HSS to resist the axial loads.

From AISC manual, Table 2-4, the material properties are as follows:

ASTM A500 Grade B

$F_y = 42$ ksi

$F_u = 58$ ksi

Determine the required strength:

The governing load combinations that include seismic effects are as follows:

LRFD load combinations (5 and 7) from ASCE 7, Section 12.4.2.3 (including the 0.5 factor on L permitted by Section 12.4.2.3)

Load combinations

5) $(1.2 + 0.2S_{DS})D + \rho Q_E + 0.5L + 0.2S$

7) $(0.9 - 0.2S_{DS})D + \rho Q_E$

The required axial compressive strength of the brace is as follows:

$P_u = [1.2 + 0.2(1.0)]14 \text{ kips} + 1.0(136 \text{ kips}) + 0.5(13 \text{ kips}) + 0.2(0)$

$= 162.1$ kips

The required axial tensile strength of the brace is as follows:

$P_u = [0.9 - 0.2(1.0)]14 \text{ kips} + 1.0(-136 \text{ kips})$

$= -162.1$ kips

The unbraced length of the brace from work point on the beam to work point on the column/base is as shown in Figure 6.8.

$L = \sqrt{(12 \text{ ft})^2 + (15 \text{ ft})^2} = 19.21$ ft

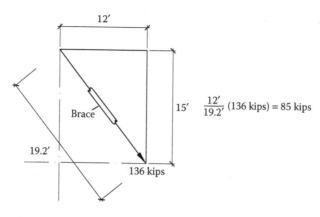

Figure 6.8 Geometry of brace from beam to column/base.

AISC seismic provisions, Section F2.4a, requires that between 30% and 70% of the total horizontal force to be resisted by braces in tension.

The total horizontal force is as follows:

57.0 kips + 54.0 kips + 37.4 kips + 20.8 kips = 169.2 kips

The horizontal component of the axial force due to earthquake force in BR-1 that is in tension is as shown in Figure 6.8.

$$\frac{85}{169.2} = 0.5(100) = 50\% \text{ of total horizontal force in line of the braced frame}$$

Therefore, it meets the AISC seismic provision.

In this example, we are selecting a brace and checking its compression and tension capacity. Try a round HSS 6.625 × 0.500 for the brace. From AISC manual Table 1-13, the geometrical properties of the pipe section are as follows:

$$D = 6.625 \quad t_{nom} = 0.500 \text{ in} \quad t_{des} = 0.465 \text{ in}$$

$$A = 9 \text{ in}^2 \quad I = 42.9 \text{ in}^2 \quad r = 2.18 \text{ in}$$

The width to thickness limitations according to AISC seismic provisions F2.5 must satisfy the requirements for highly ductile members.

Elements in the brace members must not exceed λ_{hd} width-to-thickness ratio in AISC seismic provisions (Table D1.1).

Highly ductile members for walls of round HSS.

$$\lambda_{hd} = 0.038\left(\frac{E}{F_y}\right) = 0.038\frac{29,000 \text{ ksi}}{42 \text{ ksi}} = 26.2$$

$$\frac{D}{t_{des}} = \frac{6.625}{0.465} = 14.25 < 26.2, \text{ therefore OK}$$

Since $D/t_{des} \le \lambda_{hd}$, the HSS 6.625 × 0.500 is OK for width-to-thickness limitation for highly ductile members.

Alternatively, Table 1-6 in the seismic design manual provides a simple check for members, which satisfies width-to-thickness requirements.

Table 1-6 shows that an HSS 6.625 × 0.500 used as a brace for SCBF will satisfy width-to-thickness requirements.

Brace slenderness

Use $k = 1.0$ for both the x–x and y–y axes. AISC seismic provisions F2.5b(1):

$$\frac{kl}{r} \le 200$$

$$\frac{kl}{r} = \frac{1.0(15.5 \times 12)}{2.18} = 85.3 < 200, \text{ therefore OK}$$

Second-order effects

To perform an analysis of second-order effects on the brace, use the procedure in Appendix 8 of AISC specifications—Approximate second-order analysis.

The required second-order flexural strength M_r and axial strength P_r of all members are determined as follows:

$$M_r = B_1 M_{nt} + B_2 M_{nt} \qquad \text{Appendix 8 (A–8–1)}$$

$$P_r = P_{nt} + B_2 M_{lt} \qquad \text{Appendix 8 (A–8–2)}$$

where:

B_1 is multiplier to account for $p - \delta$ effects
B_2 is multiplier to account for $P - \Delta$ effects
M_{lt} is moment due to lateral translation
M_{nt} is moment with structure restrained against translation
P_{lt} is force due to lateral translation
P_{nt} is force with structure restrained against translation

For simplicity, second-order effects will be ignored for this example.

Available compression strength

Use $L = 19.2$ ft for the unbraced length of the brace.

From AISC manual Table 4-5, the available compression strength of an HSS 6.625×0.500, with $k = 1.0$, an unbraced length $= 19.2$, $kl = 19.2$, (using interpolation) is as follows:

LRFD:

$$\phi_c P_n = 171.6 \text{ kips} > 162.1 \text{ kips, therefore OK} \qquad \text{(load case \#5)}$$

Available tensile strength

From AISC manual Table 5-6 for an HSS 6.625×0.500 brace, the available tensile yielding strength is as follows:

LRFD:

$$\phi_c P_n = 340 \text{ kips} > 126.2 \text{ kips, therefore OK} \qquad \text{(load case \#7)}$$

Therefore, a HHS 6.625×0.500 pipe section is adequate to sustain the design loads.

6.2.2 Frame analysis

The required frame analysis is based on the determination of the expected forces that the braces, which have been selected, are able to sustain in tension and compression. The brace forces are later applied to the frame's columns and beams to design those elements, which will meet the demands of the brace forces.

Step 9: Design SCBF column CL-1

In order to design column CL-1 in the frame (see Figure 6.3), we must first perform an analysis of the braces' strength in the system. An analysis of the expected strengths of the braces in tension and compression is required according to AISC seismic provisions F2.3.

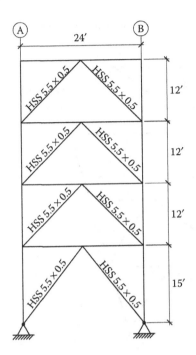

Figure 6.9 SCBF with selected brace sizes.

In this example, the braces are the same size on all floors except the main level as shown in Figure 6.9.

The selected material for all of the braces is ASTM A500, Grade B, and round HSS. Expected Strength Analysis:
ASTM A500 Grade B ·

F_y = 42 ksi

F_u = 58 ksi

From AISC seismic manual Table 1-13:
HSS 6.625 × 0.500 (main and second levels)

A = 9 in²

r = 2.18 in

HSS 5.500 × 0.500 (third and fourth levels)

A = 7.36 in²

r = 1.79 in

The size of braces determines the performance of the system in regard to ductility. Two scenarios for analysis must be examined to determine the larger forces produced between the two scenarios.

Case (i): All braces resist forces corresponding to their expected strength in compression or tension.

Lengthened brace (typ.) due to tension

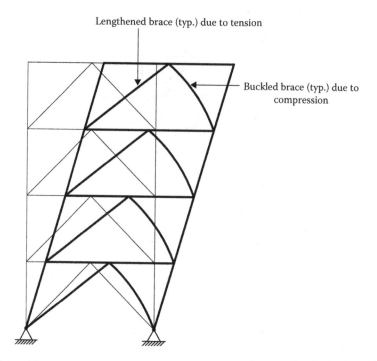

Buckled brace (typ.) due to compression

Figure 6.10 Displaced frame that produces buckled braces in compression and lengthened braces in tension.

Case (ii): All braces in tension resist forces equal to expected strength and all braces in compression resist forces equal to their postbuckling strength.

In order to do this, we have to locate the braces that will go into tension and compression when the frame moves laterally. The brace configurations—that is, an inverted V, chevron, or cross-bracing layout, will promote tension and compression in the members that lengthen and shorten, respectively (Figures 6.10 and 6.11).

The length of the braces on the main level, as previously calculated, is $L = 19.2$ ft, and the length of the braces on the second–fourth level is $L = (12/0.707) = 16.97$ ft.

From AISC seismic provision F2.3, the expected strength of the brace in tension, is found by the following equation:

$$P_{tension} = R_y F_y A_g,$$

From AISC seismic provision Table A3.1,

$$R_y = 1.4$$

The nominal compressive strength is determined as follows:

$$P_n = F_{cre} A_g \qquad\qquad \text{AISC specifications (E3–1)}$$

$R_y F_y$ is used in lieu of F_y according to the AISC seismic provisions (F2.3).

Then,

$$\frac{kl}{n_y} \leq 4.71 \sqrt{\frac{E}{R_y F_y}}$$

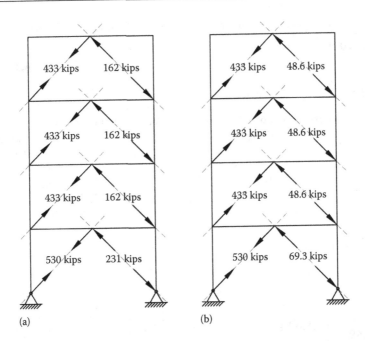

(a)   (b)

Figure 6.11 Brace forces from analysis as per seismic provisions F2.3. (a) Analysis type (i) expected strength in tension and compression. (b) Analysis type (ii) expected strength in tension and postbuckling compression.

and $F_{cre} = \left[0.658^{\frac{F_y}{F_e}} \right] F_y$ AISC specifications (E3–2)

else,

$F_{cre} = 0.877 F_e$ AISC specifications (E3–3)

where,

$F_e = \dfrac{\pi^2 E}{(kL/r)^2}$ AISC specifications (E3–4)

Check the strengths of the HSS 6.625 × 0.500:
Tension strength:

$P_{tension} = 1.4(42)9 = 529.2$ kips

Compression strength:

$\dfrac{kL}{r} = \dfrac{1.0(19.2 \times 12)}{2.18} = 105.69$

$4.71 \sqrt{\dfrac{29000}{1.4(42)}} = 104.6$

Table 6.4 Expected brace strengths in tension, compression, and postbuckling compression

Expected brace strength in tension

Brace member	A (in²)	$R_y F_y A_g$ (kips)
HSS 6.625 × 0.500	9 in²	530
HSS 5.500 × 0.500	7.36 in²	433

Expected brace strength in compression and in postbuckling compression

Brace member	Compression 1.14 $F_{cre}A_g$ (kips)	Postbuckling compression 0.3(1.14$F_{cre}A_g$)kips
HSS 6.625 × 0.500	231	69.3
HSS 5.500 × 0.500	162	48.6

Because $\dfrac{kL}{r} > 104.6$

Then,

$$F_e = \frac{\pi^2 (29,000)}{(105.69)^2} = 25.62 \text{ ksi}$$

Hence,

$$F_{cre} = 0.877(25.62) = 22.5 \text{ ksi}$$

The compression strength is then found as follows:

$$P_{compression} = 1.14\, F_{cre}A_g = 1.14(22.5)9 \text{ in}^2 = 230.85 \text{ kips}$$

The tension and compression strengths of the HSS 5.500 × 0.500 are found by the same method as calculated here (Table 6.4):

$$P_{tension} = 433 \text{ kips}$$

$$P_{compression} = 162 \text{ kips}$$

6.2.3 Column design

Now that the analysis of the braces have been performed, the forces on the columns can be developed by resolving the vertical components of the diagonal brace forces to the columns. These forces act at the joints of the intersections of the braces and the columns as well as the vertical components that are transferred to the columns by beam shear.

To perform a SCBF column design, proceed by selecting the material of the column. Here we will select an ASTM A992W-shape.

The tributary floor areas of the columns on gridlines A and B are shown in Figure 6.12 and the corresponding loading is calculated and shown in Table 6.5.

The column forces from the dead loads, live loads, and snow loads for columns on A and B lines are summed, and the seismic load shown in Figure 6.7 is included here as well and shown as reactions in Figure 6.13.

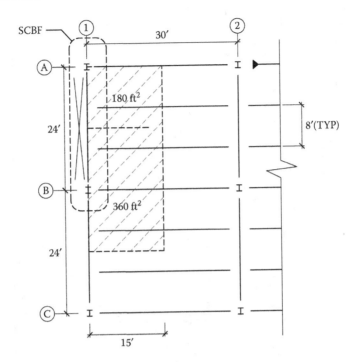

Figure 6.12 Roof and typical floor-framing plan illustrating tributary areas of frame columns.

Determine the required strength of the columns from AISC seismic provisions section F2.3 (mechanism Analysis), which requires two cases, as previously discussed, to be analyzed, and that is (i) the expected strength of the braces in compression and tension and (ii) the expected strength of the braces in tension and postbulking compression.

Distribute forces from braces shown in Figure 6.11 to the columns, vertically as shown in Figure 6.14.

The required strength of the columns according to AISC seismic provision F2.3 requires the use of the overstrength factor and as such, the load combinations provided by ASCE 7 that include the overstrength factor should be used.

From Figure 6.14, the amplified seismic load on the column in tension and compression is as follows:

$$P_{E_{mh}} = 728 \text{ kips} \quad \text{(compression)}$$

$$T_{E_{mh}} = -534 \text{ kips} \quad \text{(tension)}$$

Substitute $P_{E_{mh}}$ and $T_{E_{mh}}$ for the overstrength factor seismic force $\Omega_0 Q_E$ in the seismic load combinations from AISC 7.

We then have, for column B:

Load case 5: $P_a = (1.2 + 0.2S_{DS})P_D + P_{E_{mh}} + 0.5P_L + 0.2P_S$

$$= \left[1.2 + 0.2(1.0)\right](134.28) + 728 + 0.5(86.4) + 0.2(7.2)$$

$$= 961 \text{ kips}$$

Table 6.5 Column A and B loadings per floor

Level		Column A				Column B		
		Dead load (k)	Live load (k)	Snow load (k)		Dead load (k)	Live load (k)	Snow load (k)
Roof	D_r	0.070(180) = 12.6				0.070(360) = 25.2		
	c.w$_r$	(12 + 15)(0.120) = 3.24				24(0.120) = 2.88		
	S			0.020(180) = 3.6				0.020(360) = 7.2
4th	D_f	0.085(180) = 15.3				0.085(360) = 30.6		
	c.w.	(12 + 15)(0.200) = 5.4				24(0.200) = 4.8		
	L_f		0.080(180) = 14.4				0.080(360) = 28.8	
3rd	D_f	15.3				30.6		
	c.w.	5.4				4.8		
	L_f		14.4				28.8	
2nd	D_f	15.3				30.6		
	c.w.	5.4				4.8		
	L_f		14.4				28.8	
		77.94 43.2 3.6				134.28 86.4 7.2		

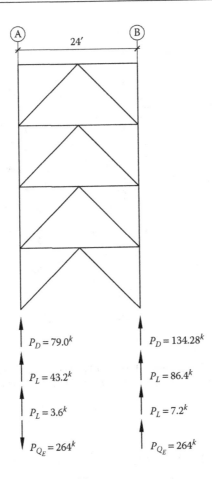

Figure 6.13 Vertical reactions at columns A and B.

and

Load case 7: $P_u = (0.9 - 0.2S_{DS})P_D + T_{Emh}$

$$= [0.9 - 0.2(1.0)](134.28) + (-534)$$

$$= -440 \text{ kips}$$

The required strength of the column does not need to exceed the forces determined using load combinations stipulated by the applicable building code as was previously performed and is shown in Figure 6.7.

Therefore, check the following load combinations:

5) $P_u = (1.2 + 0.2S_{DS})P_D + \Omega_0 Q_E + 0.5P_L + 0.2P_S$

$$= [1.2 + 0.2(1.0)](134.28) + 2.0(264) + 0.5(86.4) + 0.2(7.2)$$

$$= 760 \text{ kips}$$

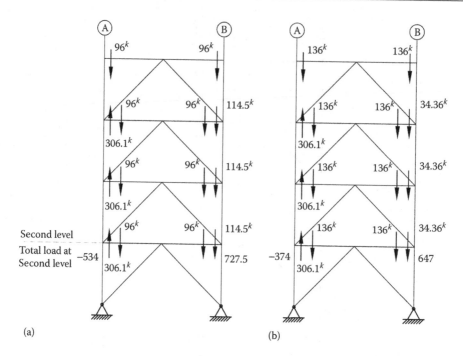

Figure 6.14 Forces from brace analyses imposed upon columns. (a) Expected strength in tension and compression and (b) expected strength in tension and postbuckling compression.

and,

7) $P_u = (0.9 - 0.2 S_{DS})P_D + \Omega_0 Q_E + 1.6 P_H$

$= [0.9 - 0.2(1.0)](134.28) + 2.0(-264)$

$= -434 \text{ kips}$

Required axial strength of column CL-1 on gridline B (Table 6.6):

Second-order effects

Because the governing loads are code-based instead of the mechanism analysis (expected force based on capacity of the braces), second-order effects should be considered. However, for this example the second-order effect will be ignored for simplicity.

Table 6.6 Expected strengths as per seismic provisions F2.3

Analysis Type: AISC Seismic Provisions F2.3			
Expected strengths in tension and compression		Code-specified amplified seismic loads (use of overstrength factor)	
Compression	Tension	Compression	Tension
961 kips	−440 kips	760 kips (govern)	−434 kips (govern)

Continue by selecting a W12 column (based on compression strength):
From AISC manual Table 4-1,
  Use $k = 1.0$ (for a pin–pin condition)
  $kl = 15$ (unbraced length)

W12 × 79

$\phi_c P_n = 809 > 760$ kips, therefore OK

Check tension strength
From AISC manual Table 5-1,

W12 × 79

$\phi_t P_n = 1040 > 434$ kips, therefore OK

Check width-to-thickness limits
SDM (seismic design manual), Table 1-3
W12 × 79 does not satisfy the seismic requirements, therefore select the next size in the table.
A W12 × 96 column does satisy width-to-thickness limits
Therefore, use W12 × 96 Column.

6.2.4 Beam design

Similar to the column design of the frame, an analysis of the loading on the beam due to the dead load and live load transferred from the floor framing must be done. The beam is conservatively analyzed as a simple beam to determine shears and moments, and then the vertical components from the braces are determined and applied to the point of connection of the braces on the beam. The design procedure is outlined as follows:

Step 10: Design SCBF Beam BM-1
  Select an ASTM A992 W-shape. Design the beam as a composite beam for strength (braced by the floor slab).
  Assume the beam to be simply supported and calculate the beam shears and moments as shown in Figure 6.15.

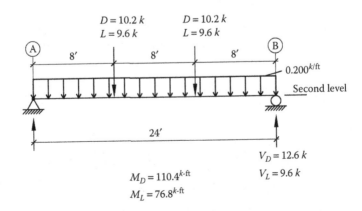

Figure 6.15 Free body diagram of BM-1 conservatively modeled as a simply supported beam.

Material properties:
ASTM A992

$F_y = 50$ ksi

$F_u = 65$ ksi

As required by AISC seismic provisions F2.3, the amplified seismic load is determined from the larger of the two analyses.
   (i)   Expected strength in compression or in tension
   (ii)   Expected strength in tension and expected postbuckling strength
The forces from the analyses are noted in Table 6.5 and illustrated for the whole frame in Figure 6.11 and are also shown in Figure 6.16 for convenience.
For analysis type (i) the unbalanced vertical force is determined from the vertical component of the brace force on the beam.
Analysis (i), the vertical force on the beam (Figure 6.17)

$P_y = (231 - 530)\sin 51.34° = -233.5$

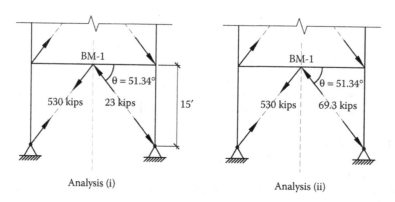

Analysis (i)   Analysis (ii)

Figure 6.16 Expected strengths of braces.

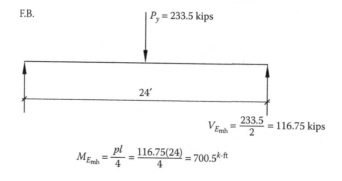

F.B. $P_y = 233.5$ kips

24′

$V_{E_{mh}} = \dfrac{233.5}{2} = 116.75$ kips

$M_{E_{mh}} = \dfrac{pl}{4} = \dfrac{116.75(24)}{4} = 700.5^{k\text{-ft}}$

Figure 6.17 Moment induced by the strengths of the braces (case i).

Calculate the horizontal force in the beam, which is the difference between the sums of the two force components from the braces. Assume the horizontal force is shared evenly between the floors below and above.

$$P_x = \frac{\Sigma \text{horizontal component of braces}}{2} = \frac{(231-530)\cos 51.3°}{2} = -187 \text{ kips}$$

P_x is assumed to act in either direction.
Using the following load combinations
Analysis (i)
Load case (5) axial force

$$P_u = (1.2 + 0.2S_{DS})V_D + P_{Emh} + 0.5V_L + 0.2V_S$$

$$= [1.2 + 0.2(1.0)](0) + 187 + 0.5(0) + 0.2(0)$$

$$= 187 \text{ kips}$$

Load case (5) shear force

$$V_u = (1.2 + 0.2S_{DS})V_D + V_{Emh} + 0.5V_L + 0.2V_S$$

$$= [1.2 + 0.2(1.0)](12.6) + 116.75 + 0.5(9.6) + 0.2(0)$$

$$= 139.2 \text{ kips}$$

Load case (5) moments

$$M_u = (1.2 + 0.2S_{DS})M_D + M_{Emh} + 0.5M_L + 0.2M_S$$

$$= [1.2 + 0.2(1.0)]110.4 + 700.5 + 0.5(76.8) + 0.2(0)$$

$$= 893.5 k - ft$$

Analysis (ii), the vertical force on the beam (Figure 6.18)

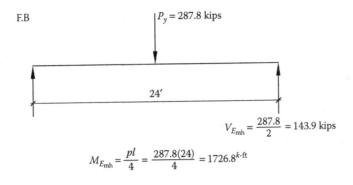

$$V_{Emh} = \frac{287.8}{2} = 143.9 \text{ kips}$$

$$M_{Emh} = \frac{pl}{4} = \frac{287.8(24)}{4} = 1726.8^{k-ft}$$

Figure 6.18 Moment induced by the strengths of the braces (case ii).

$P_y = (-530 + 69.3)\sin 51.34° = -287.8$ kips

$$P_x = \frac{\Sigma \text{horizontal component of braces}}{2} = \frac{(-530 + 69.3)\cos 51.3°}{2} = -144 \text{ kips}$$

Using the following load combinations
Analysis (ii)
Load case (5) axial force:

$P_u = (1.2 + 0.2S_{DS})V_D + P_{E_{mh}} + 0.5V_L + 0.2V_S$

$= [1.2 + 0.2(1.0)](0) + 144 + 0.5(0) + 0.2(0)$

$= 144$ kips

Load case (5) shear force:

$V_u = (1.2 + 0.2S_{DS})V_D + V_{E_{mh}} + 0.5V_L + 0.2V_S$

$= [1.2 + 0.2(1.0)](12.6) + 143.9 + 0.5(9.6) + 0.2(0)$

$= 166.34$ kips

Load case (5) moments:

$M_u = (1.2 + 0.2S_{DS})M_D + M_{E_{mh}} + 0.5M_L + 0.2M_S$

$= [1.2 + 0.2(1.0)]110.4 + 1726.8 + 0.5(76.8) + 0.2(0)$

$= 1919.8k - ft$

Summary of the required strengths
Analysis (i)

$P_u = 187.0$ kips

$V_u = 139.2$ kips

$M_u = 893.5k - ft$

Analysis (ii)

$P_u = 144.0$ kips

$V_u = 166.34$ kips

$M_u = 1919.8k - ft$

Then the available flexural strength is calculated as follows:

The beam bracing is provided at 8′ intervals (see the floor-framing plan). However, the floor slab can be considered to continuously brace the beam and therefore lateral torsional buckling does not apply and the available flexural strength is based on the plastic moment of the beam.
Select a beam from AISC manual Table 3-6 based on flexural strength.
W27×178 Table 3-6 (fully braced)

$r_x = 11.6$

$A_g = 52.5$

$I = 7020 \text{ in}^4$

$\phi_b M_p = 2140^{k \cdot f} > 1919.8^{k \cdot f}$, therefore OK

Check slenderness of the W27 × 178
AISC Seismic Design Manual
Table 1-3
W27 × 178
SCBF
OK for beam
Available compression strength:
    In compression, the beam is considered to be continuously braced by the slab, so minoraxis flexural buckling does not apply (about the y–y axis). The beam is assumed unbraced about the major axis (x–x axis) for the full length of the beam, $kl = 24$ ft.
Determine the compression strength of the W27 × 178.
Unbraced length.

$$\frac{kl}{r_x} = \frac{1.0(24 \times 12)}{11.6} = 24.83$$

AISC Manual.
Table 4-22.

$$\frac{kl}{r_x} = 25, \; F_y = 50 \text{ ksi}$$

$\phi_c F_{cv} = 43 \text{ ksi}$

$\phi_c F_n = \phi_c F_{cv} A_g = 43(52.5 \text{ in}^2) = 2257.5 \text{ kips}$

Consider second-order effects (moment amplification)
No translation, $B_2 = 0$
The Euler buckling load

$$P_{e1} = \frac{\pi^2 EI}{(kl)^2}$$

$$= \frac{\pi^2 (29,000 \text{ ksi})(7020 \text{ in}^4)}{(1.0 \times 40 \times 12)^2}$$

$$= 8721 \text{ kips}$$

$$B_1 = \frac{C_m}{1 - [\alpha P_n/P_{e1}]} \leq 1.0 \qquad\qquad\qquad \text{(AISC spec. C2–2)}$$

$\alpha = 1.0$ for LRFD

$C_m = 1.0$ conservatively

$$B_1 = \frac{1.0}{1 - [1.0 \times 187/8721]} = 1.022$$

Analysis (i) with second-order effects

$$M_u = B_1(1.2 + 0.2S_{DS})M_D + M_{E_{mh}} + B_1(0.5)M_L + 0.2M_S$$

$$= 1.022(1.4(1.0))(110.4) + 700.5 + 1.022(0.5)76.8$$

$$= 897.7k - \text{ft}$$

Analysis (ii) with second-order effects

$$M_u = B_1(1.2 + 0.2S_{DS})M_D + M_{E_{mh}} + B_1(0.5)M_L + 0.2M_S$$

$$= 1.022(1.4(1.0))(110.4) + 1,726.8 + 1.022(0.5)76.8$$

$$= 1,924.0k - \text{ft}$$

Summary of forces analysis (i)

$P_u = 187.0$ kips

$V_u = 139.2$ kips

$M_u = 897.7k - \text{ft}$

Summary of force analysis (ii)

$P_u = 144.0$ kips

$V_u = 166.34$ kips

$M_u = 1924.0k - \text{ft}$

Combined loading for axial and flexural forces:
Analysis (i)

$$\frac{P_n}{P_c} = \frac{187 \text{ kips}}{2257.5 \text{ kips}} = 0.083$$

Because $\frac{P_n}{P_c} \leq 0.2$, the beam column design is controlled by equation

$$\frac{P_r}{2P_c} + \left[\frac{M_{nx}}{M_{cx}} + \frac{M_{ny}}{M_{cy}} \right] \leq 1.0 \qquad\qquad \text{(AISC Spec. Equation H1–1b)}$$

$$\frac{187}{2(2257.5)} + \left[\frac{897 \text{ kip} - \text{ft}}{2140 \text{ kip} - \text{ft}} + 0 \right] = 0.46 \leq 1.0, \text{ therefore OK}$$

Analysis (ii)

$$\frac{P_n}{P_c} = \frac{144 \text{ kips}}{2257.5 \text{ kips}} = 0.06 < 0.2$$

Then,

$$\frac{P_r}{2P_c} + \left[\frac{M_{nx}}{M_{cx}} + \frac{M_{ny}}{M_{cy}} \right] \leq 1.0$$

$$\frac{144}{2(2257.5)} + \left[\frac{1924 \text{ kip} - \text{ft}}{2140 \text{ kip} - \text{ft}} + 0 \right] = 0.93 \leq 1.0, \text{ therefore OK}$$

Check shear strength of $W27 \times 178$:
    From AISC Manual Table 3-2,

$$\phi_v V_n = 605 \text{ kips} > 166.34 \text{ kips, therefore OK}$$

Therefore, use $W27 \times 178$ for SCBF beam BM-1.

Chapter 7

Concrete

7.1 INTRODUCTION TO LATERAL CONCRETE DESIGN

7.1.1 Introduction and general information

The design standard for structural concrete is the Building Code Requirements for Structural Concrete known as ACI-318 and associated commentary. The International Building Code (IBC) is the governing general building code whereby Chapter 16 prescribes the required load combinations and references the use of ASCE 7 for wind and seismic loading to be used for the design of concrete structures. Chapter 19 of the IBC codifies the provisions for concrete, which govern the materials, quality control, design, and construction of concrete used in structures. The IBC references ACI 318 as the accepted standard for design accounting for the appropriate amendments and modifications noted in Chapter 19.

7.1.2 Design methods

The design of concrete is based on proportioning for adequate strength using load factors and strength-reduction factors. All members of frames or continuous construction are to be designed to the maximum effects of factored loads based on elastic analysis; however, ACI allows to simplify design, where appropriate by using assumptions.

As an alternate to frame analysis, the following approximate moments and shears are permitted for the design of continuous beams and one-way slabs (slabs reinforced to resist flexural stresses in only one direction), provided (1) thorough (5) are satisfied:

1. There are two or more spans
2. Spans are approximately equal, with the larger of two adjacent spans not greater than the shorter by more than 20%
3. Loads are uniformly distributed
4. Unfactored live load, L, does not exceed three times unfactored dead load, D
5. Members are prismatic

For calculating negative moments, l_n is taken as the average of the adjacent clear span lengths.

Positive moment
    End spans
        Discontinuous end unrestrained $w_u l_n^2/11$
        Discontinuous end integral with support $w_u l_n^2/14$
        Interior spans $w_u l_n^2/16$

Negative moments at the exterior face of first interior support

Two spans $w_u l_n^2/9$

More than two spans $w_u l_n^2/10$

Negative moment at other faces of interior supports $w_u l_n^2/11$

Negative moment at the face of all supports for slabs with spans not exceeding 10 ft; and beams where the ratio of sum of column stiffness to beam stiffness exceeds 8 at each end of the span $w_u l_n^2/12$

Negative moment at the interior face of exterior support for members built integrally with supports

Where support is spandrel beam $w_u l_n^2/24$

Where support is a column $w_u l_n^2/16$

Shear in end members at the face of first interior support $1.15 w_u l_n/2$

Shear at the face of all other supports $w_u l_n/2$

7.1.3 Lateral concrete systems

Concrete lateral systems are essentially of two types; shear wall or moment frame systems. The systems are classified as ordinary, intermediate, or special based on ductility requirements associated seismic design category (SDC).

As discussed in the previous chapters, the seismic loads, which a building's structural system is to resist, are inertial forces, and they are a function of the building's mass. The goal of the designer of the structural system is to proportion the building such that the building's structural system responds to moderate to high seismic events nonlinearly so the building's composition can take advantage of ductile performance in order to absorb forces due to lateral movements. A system's ductility is built into the design by the seismic response factor or coefficient R. The higher the R, the more ductile the system is. The load path of a structural system is extremely important as we discussed when reviewing horizontal and vertical irregularities in Chapter 5, such that it defines the demands of the system. Consequently, seismic design as presented in the ACI concrete design code is based on the ductile design requirements of Chapters 1 through 18 and Chapter 21—Earthquake Resistant Structure.

Also, as previously discussed, wind loads on buildings are external loads and designing for wind loads is based on linear behavior. The ACI code requirements referencing linear of structural behavior in Chapters 1 through 18 apply.

It follows, those structural systems having higher forces or effects due to wind loading than seismic forces due to seismic loading are still required to meet ductile detail requirements due to seismic design.

The procedure for design is as follows:

1. Calculation and determination of wind loading on the building for the main wind force-resisting system as well as for components and cladding
2. Determine the SDC of the building based on structure type and risk category and ground accelerations specific to the site, which sets design and detailing requirements
3. Determine the seismic design forces as described in Chapter 5, which includes calculating the base shear, vertical distribution of base shear, determining redundancy, horizontal distribution of forces to lateral resisting elements

7.1.4 Development length of reinforcing to meet seismic ductile requirements

Development length for standard 90° hooked bars ACI 318 21.7.5.1
(For No. 3 through No. 11 bars)

$$l_{dn} = \frac{F_y d_b}{65\sqrt{f_c'}}$$

For light weight concrete, use 25% length increase.
Development length for straight bars in tension ACI 318 21.7.5.2
(For No. 3 through No. 11 bars)

$$l_d > 2.5 l_{d_b}$$

When the depth of concrete below the bar does not exceed 12" (cast in one lift)

$$l_d > 3.25 l_{d_b}$$

When the depth of concrete below the bar exceed 12" (cast in one lift).
For nonseismic development length, see ACI 318 12.2.
Seismic lap spliced:

Tension lap splices ACI 318 12.15
   Class A splice $1.0 l_d$
   Class B splice $1.3 l_d$
   l_d is as presented in ACI 12.2.2
   Use class B lap splices for seismic detailing.
Compression lap splices ACI 318 12.16
   For $f_y \ll 60,000$ psi: lap splice length $= 0.0005 f_y d_b$
   For $f_y > 60,000$ psi: lap splice length $= (0.0009 f_y - 24) d_b$

$$l_{d_{min}} \geq 12''$$

and for $f_c' < 3,000$ psi: $l_d = 1.3 l_d$
No splices are permitted in region of plastic hinges.

7.2 SHEAR WALL SYSTEMS

Shear walls are defined by the SDC and required detailing as shown in Table 7.1 below.

Table 7.1 Shear wall design requirements

Wall type	Design requirements	SDC
Ordinary reinforced concrete shear wall	AC I, Chapters 1–18	A, B, C
Special reinforced concrete shear walls	AC I, Chapters 1–18 and Section 21.1.3 to 21.1.7 and 21.9	A, B, C, D, and F

Example 7.1: Special reinforced concrete shear wall design

This example illustrates the design of a special reinforced concrete shear wall (Figure 7.1).

The concrete wall is 12 inches thick, and it is part of the seismic lateral load-resisting system of a building located in an SDC D site, the following information is given (Figures 7.2 and 7.3):

$\rho = 1.0$ $F_c' = 4000\,\text{psi}$

$S_{DS} = 1.0$ $F_y = 60,000\,\text{psi}$

$I_e = 1.0$

The shear wall is required to be design as special.
Gravity loads (including wall weights)

$D_R = 20\,\text{kips}$

$L_R = 10\,\text{kips}$

$D_{\text{floor}} = 30\,\text{kips}$

$L_{\text{floor}} = 40\,\text{kips}$

Load combinations (Table 7.2)

1. $(1.2 + 0.2S_{DS})D + \rho Q_E + 0.5L$

2. $(0.9 - 0.2S_{DS})D + \rho Q_E$

1. $P_u = \left[1.2 + 0.2(1.0)\right](30 + 30 + 20) + 0.5(40 + 40 + 10) = 112 + 45 = 157\,\text{kips}$

2. $P_u = \left[0.9 - 0.2(1.0)\right](30 + 30 + 20) = 56\,\text{kips}$

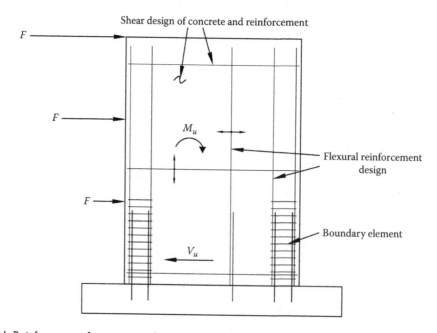

Shear design of concrete and reinforcement

M_u

Flexural reinforcement design

F

V_u

Boundary element

Figure 7.1 Reinforcement for concrete shear wall.

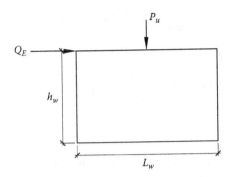

Figure 7.2 Free-body diagram of the lateral seismic and vertical loading on a wall.

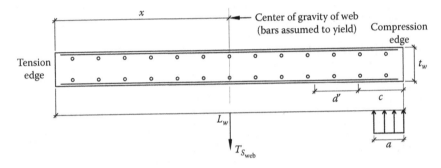

Figure 7.3 Plan section of the shear wall illustrating the anticipated compression and tension zones of the wall.

Table 7.2 Tabulated design axial loads, shear forces, and moments

| | | Main level | |
Load case	P_u (k)	V_u (k)	M_u (k – ft)
1	157	289	7249
2	56	289	7249

1. $V_u = \rho Q_E = 1.0(112 + 96 + 81) = 289 \, \text{kips}$

2. $V_u = \rho Q_E = 1.0(112 + 96 + 81) = 289 \, \text{kips}$

1. $M_u = 112(40) + 96(28) + 81(16) = 7,249^{k \cdot ft}$

2. $M_u = 112(40) + 96(28) + 81(16) = 7,249^{k \cdot ft}$

Shear design:

The distributed web reinforcement ratios of $\rho = 0.0025$ for the longitudinal and transverse web reinforcement are required if V_u exceeds the following equation:

$$A_{cv} \lambda \sqrt{F_c'} \qquad \text{ACI 21.9.2}$$

$\lambda = 1.0$ for normal weight concrete

A_{cv} = area of concrete wall subject to shear

$$A_{cv}\lambda\sqrt{F_c'} = \frac{(12''\text{ thick wall})(18'\times12''/\text{ft})1.0\sqrt{4000}}{1000^{\#/k}} = 164\,\text{kips}$$

$$V_u = 289 > 164$$

Therefore, the minimum web reinforcement is required.
  Wall reinforcement:

$$\rho = 0.0025(12''\times12'') = 0.36^{\text{in}^2/\text{ft}}$$

If $t_w > 10''$ provide two curtains of reinforcement ACI 14.3.4
  Use #4 @ 12" O.C. Each face

$$A_{s_t} = 0.2\,\text{in}^2\times2\,\text{layers} = 0.4\,\text{in}^2 > 0.36\,\text{in}^2, \text{ therefore OK}$$

Shear strength of special shear walls:

$$\phi V_n = \phi A_{cv}(\alpha_c\lambda\sqrt{F_c'} + \rho F_y)$$

$$\phi = 0.6$$

$$\frac{h_w}{L_w} = \frac{40}{18} = 2.22 > 2.0$$

Thus, $\alpha_c = 2.0$

$$\rho_t = \frac{0.4\,\text{in}^2}{144\,\text{in}^2} = 0.0028$$

$$\phi V_n = 0.6(12\times216)\frac{\left[2.0(1.0)\sqrt{4000} + 0.0028(60,000)\right]}{1000} = 458^{\text{kips}} > 289^{\text{kips}}, \text{ therefore OK}$$

Flexural design of wall
  Minimum distributed vertical reinforcement in wall (Figure 7.4)

$$\rho_l = 0.0025:$$

$$\rho_{sl} = \rho(bd) = 0.0025(12''\times12'') = 0.36^{\text{in}^2/\text{ft}}$$

Use #4 bars @12" O.C. (same as transverse steel A_{s_t})

$$A_{s_{web}} = \text{No. of end bars} + \text{longitudinal bars in web}$$

$$= 4 - \#10\text{ bars} + 34 - \#5\text{ bars}$$

$$= 4(1.27) + 34(0.31) = 15.62\,\text{in}^2$$

$$a = \frac{A_{s_{web}}F_y + P_u}{0.85F_c't_w}$$

P_u = use load case #2, the least axial load from both load cases.

$$= \frac{15.62\,\text{in}^2(60\,\text{ksi})}{0.85(4\,\text{ksi})(12\,\text{in})} = 22.97\,\text{in}$$

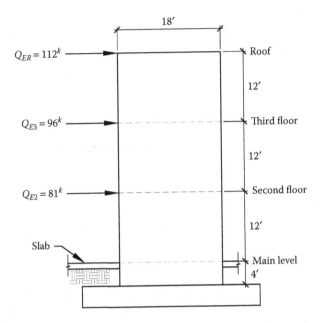

Figure 7.4 Seismic story forces applied to the special reinforced concrete shear wall.

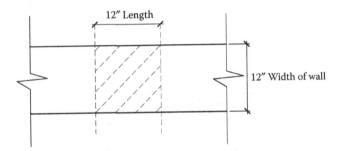

Figure 7.5 Cross-sectional area of the wall to determine minimum web reinforcement.

Check if boundary reinforcement is required.

If $\phi M_{n_{web}} + P_u \leq M_u$ then boundary reinforcement is required (Figure 7.5).
The d of the wall is estimated at half the length of the wall conservatively.

$$d = \frac{18' \times 12''/\text{ft}}{2} = 108''$$

$$\phi M_n = 0.9 A_{s_w} F_y \left(d - \frac{a}{2} \right)$$

$$= 0.9(15.62)60\,\text{ksi}\left(108 - \frac{22.97}{2} \right)$$

$$= 81,408.5^{k \cdot \text{in}} = 6784^{k \cdot \text{ft}}$$

$$\phi M_n < M_u = 7249$$

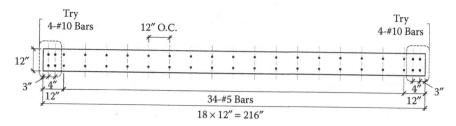

Figure 7.6 Section of wall with boundary elements included at the end of the wall.

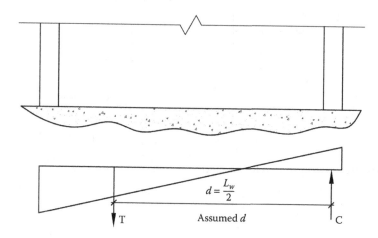

Figure 7.7 Simplified balance of forces.

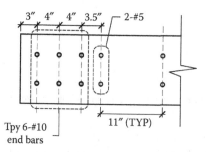

Figure 7.8 Adjusted boundary element section.

Because the design moment is less than the ultimate moment, which is needed to sustain, we must include boundary reinforcement at ends of wall (Figures 7.6 and 7.7).

Adjust spacing of bars for new reinforcement trial (Figure 7.8):

$$216'' - 2(14.5) = 187''$$

$$\text{Spacing of \#5 bars} = \frac{187''}{11''} = 17 \text{ spaces}$$

Adjust compression block based on new end bars (Figures 7.9 through 7.12).

$$A_{sweb} = 6 - \#10 \text{ bars} + 36 - \#5 \text{ bars}$$

$$= 6(1.27 \text{ in}^2) + 36(0.31 \text{ in}^2) = 18.78 \text{ in}^2$$

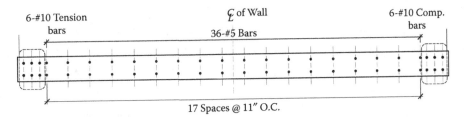

6-#10 Tension bars

₵ of Wall

36-#5 Bars

6-#10 Comp. bars

17 Spaces @ 11" O.C.

Figure 7.9 Adjusted wall section showing new boundary elements and web reinforcement.

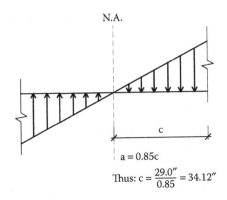

N.A.

c

$a = 0.85c$

Thus: $c = \dfrac{29.0''}{0.85} = 34.12''$

Figure 7.10 Location of the neutral axis.

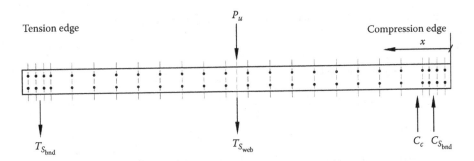

P_u

Tension edge

Compression edge

x

$T_{S_{bnd}}$ $T_{S_{web}}$ $C_c \; C_{S_{bnd}}$

Figure 7.11 Free-body diagram of the shear wall.

$$a = \frac{A_{s_{web}}F_y + P_u}{0.85F_c't_w} = \frac{18.78(60) + 56}{0.85(4)(12)} = 29.0 \, \text{in}$$

$$\epsilon_s = \frac{F_y}{E} = \frac{60,000}{29,000,000} = 0.00207 \; \text{yield}$$

Compression

⑤ $\dfrac{\epsilon_s'}{(34.12 - 3)} = \dfrac{0.003}{34.12} \Rightarrow \epsilon_s' = 0.0027 \; \text{yield}$

④ $\Rightarrow \epsilon_s' = 0.0024 \; \text{yield}$

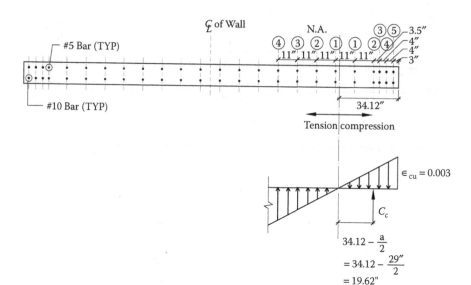

Figure 7.12 Analysis of reinforcing bars yielding in compression and tension.

③ $\Rightarrow \epsilon_s' = 0.00203$ not yielding, $F_y = 0.00203(29,000) = 58.95$ ksi

② $\Rightarrow \epsilon_s' = 0.00173$ not yielding, $F_y = 0.00173(29,000) = 50.03$ ksi

① $\Rightarrow \epsilon_s' = 0.000758$ not yielding, $F_y = 0.000758(29,000) = 21.98$ ksi

Tension

① $\Rightarrow \epsilon_s' = 0.000209$ yield

$C_c = -C_{sbnd} + T_{sweb} + T_{sbnd} + P_u = 683.62$

$M_n = 9876^{k \cdot ft}$

$\phi M_n = 0.9(9,876) = 8888^{k \cdot ft} > 7249^{k \cdot ft}$ OK satisfactory design

Boundary elements of special structural walls ACI 21.9.6 (Table 7.3)
The boundary element requirement is dictated by the largest depth of neutral axis.
Use P_u from load case 1

$P_u = 157^k$

Then,

$$C_{crequired} = -C_{sbnd} + T_{sweb} + T_{sbnd} + P_u$$

$$= -499.18 + 669.6 + 457.2 + 157$$

$$= 784.62 \text{ kips}$$

Table 7.3 Tabulated forces on the shear wall

Force component		A_s(in²)	C or T (k)	X(in)	$\in_s(x)$ or $\in'_s(x)$ (k–ft)
			Wall flexural capacity		**X = distance from N.A.**
$C_{S_{bnd}}$	⑤	2.54 (yield)	2.54(60) = 152.4	31.12	−396.24
	④	2.54 (yield)	2.54(60) = 152.4	27.12	−344.42
	③	2.54	2.54(58.95) = 149.73	23.12	−288.48
	②	0.62	0.62(50.03) = 31.02	19.62	−50.72
	①	0.62	0.62(21.98) = 13.63	8.62	−9.79
			499.18		−1089.65
$T_{S_{web}}$		36–#5 bars =11.16 (yields) 11.16 (yield)	11.16(60) = 669.6	$\dfrac{216-34.12}{2}=91''$	5074.45
$T_{S_{bnd}}$		2.54 (yield)	2.54(60) = 152.4	178.9	2272.03
		2.54 (yield)	2.54(60) = 152.4	174.9	2221.23
		2.54 (yield)	2.54(60) = 152.4	170.9	2170.43
P_u			56	73.88	344.71
C_c			683.86	19.62	−1117.72
					9875.54

To find the depth to N.A.:

$$a = \frac{C_c}{0.85f'_c b_w} = \frac{784.62}{0.85(4)(12)} = 19.23''$$

$$a = 0.85c \Rightarrow c = \frac{a}{0.85} = 22.62 \text{ in.}$$

Compression zones are to be reinforced with special boundary elements where (Figure 7.13):

$$c \geq \frac{l_w}{600(\delta_n/h_w)}$$

where:
   δ_n is design displacement (in)
   and h_w is height of wall (in)

in no case shall (δ_n/h_w) be less than 0.007

$$c = 22.62 < \frac{216}{600\left[6''/(40 \times 12)\right]} = 28.80 \text{ in.}$$

Therefore, no special boundary elements required.

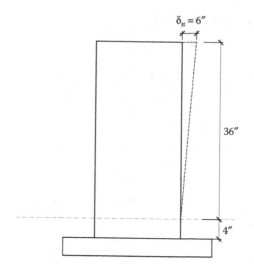

$\delta_n = 6''$

36"

4"

Figure 7.13 Deflection at the top of the wall determined by separate analysis.

Check longitudinal reinforcement at the wall boundary ACI 21.9.6.5

$$\rho = \frac{A_s}{a(t_w)} = \frac{6 \times 1.27}{22.62(12'')} = 0.028$$

$$> \frac{400}{F_y} = \frac{400}{60,000} = 0.007$$

Since $\rho > \dfrac{400}{F_y}$

Provide confinement to boundary reinforcement ACI 21.6.4.2 and 21.9.6.4(a)
Boundary element shall extend the greater of the following (Figure 7.14):

$$c - 0.1 l_w = 22.62 - 0.1(216) = 1.02 \text{ in.}$$

$$\frac{c}{2} = \frac{22.62}{2} = 11.31 \text{ in.} \leftarrow \text{controls}$$

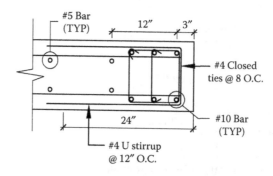

#5 Bar
(TYP)   12"   3"

#4 Closed
ties @ 8 O.C.

24"   #10 Bar
                       (TYP)

#4 U stirrup
@ 12" O.C.

Figure 7.14 Detail of the shear wall boundary element and required confinement reinforcing.

7.3 MOMENT FRAME SYSTEMS

Moment frames resist lateral forces by the strength of their joints. That is, their joints remain rigid or fixed at 90° from each other, when laterally loaded and by doing such, cause the beams and columns to bend. The member's ability to resist bending gives the frame its strength. Naturally, as the joints rotate within the frame, the beams and columns, at the joints of the frame, develop moments and shears. The objective of a successful design is to limit damage appropriately by detailing the members and joints, so a strong-column weak-beam configuration in regard to strength is developed. The demands on the frame come from both gravity and lateral loads and the applicable load combinations must be used to determine the most critical governing effects for which design is based (Figure 7.15).

7.3.1 Ordinary moment frames ACI 21.2

- Design for factored loads using provisions of ACI Chapters 1 through 18
- No special detailing requirements, except for SDCB

Beams: provide at least two of the longitudinal bars continuous along the top and bottom faces. Continue through columns or develop bars at the face of support.

Columns: columns having clear height less than or equal to five times the dimension $C_1.C_1$ = dimension of column in the direction of the span for which moments are being determined.

7.3.2 Intermediate moment frames ACI 21.3

The detail requirements of intermediate moment frames, as compared to ordinary moment frames, are intended to reduce the risk of shear failure.

Design for factored loads using all provisions of Chapter 1 through 18 and Section 21.3 (Table 7.4).

Shear strength:

Beams: ϕV_n shall not be less than the smaller of the following:

- The sum of the shear associated with the nominal moment of the beam and the shear calculated for gravity loads

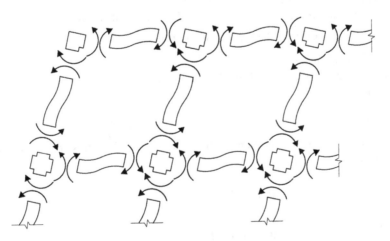

Figure 7.15 Diagrammatic representation of the moments developed due to lateral forces.

Table 7.4 Reinforcement for concrete moment frames

Frame type	Design requirements	SDC
Ordinary moment frame	ACI 318 Chapter 1–18 and Section 2.2	A, B
Intermediate moment frame	ACI 318 Chapter 1–18 and Section 21.3	A, B, C
Special moment frame	ACI 318 Chapter 1–18 and Section 21.5–21.8	A, B, C, D, E, and F

- Two times the maximum shear obtained from load combinations that include seismic loads

Columns: ϕV_n of columns shall not be less than the smaller of the following:

- The shear associated with the nominal moment of the column
- The maximum shear obtained from design load combinations that include seismic loads

Beam design ACI 21.3.4:

The positive moment strength at the face of the joint shall not be less than 1/3 the negative moment strength at the face of the joint.

- At both ends of the beam, hoops shall be provided for a length equal to two times the depth of the beam, measured from the face of support
- Spacing of the hoops shall not exceed the smaller of the following:
  - $d/4$
  - Eight times the smallest longitudinal
  - 24 times the diameter of the hoops
  - 12 inches

Column design ACI 21.3.5.1:

Columns shall be spirally reinforced in accordance with ACI 7.10.4 or shall conform with ACI 21.3.5.2-21.3.5.4

Hoops are summarized here as follows:

Hoops to be provided over a length l_0 and provided at spacing s_0.
s_0 shall not exceed the smaller of the following:

- Eight times the diameter of the smallest longitudinal bar
- 24 times the diameter of the hoop bar
- One-half the smallest cross-sectional dimension of the column

and l_0 shall not be less than the largest of the following:

- 1/6 of the clear span of the column
- Maximum cross-sectional dimension of the column
- 18 inches

7.3.3 Special moment frames

Seismic provisions for special moment frame beams. ACI 21.5 flexural members of special moment frames

$$P_u \leq \frac{A_g f_c'}{10} \qquad \text{ACI 21.5.1.1}$$

$$\frac{\text{clean span}}{d} \geq 4 \qquad \text{ACI 21.5.1.2}$$

- Width b_w not less than 0.3 h or 10 inches

$A_s \geq \dfrac{3\sqrt{f'}}{f_y} b_w d$, applies to top and bottom reinforcement   ACI 21.5.2.1

- $\rho_{max} = 0.25$, where $\rho = \dfrac{A_s}{b_w d}$

- Positive moment strength at joint face shall not be less than 1/2 the negative moment strength provided at face of joint

Shear strength requirements (Figure 7.16):
  Design force ACI 21.5.4.1

$$V_e = \dfrac{M_{pr1} + M_{pr2}}{l_n} \pm \dfrac{w_n l_n}{2}$$

Where end moments, M_{pr}, are based on steel tensile stress of $1.25 f_y$,

l_n = clean span.

Transverse reinforcement ACI 21.5.4.2

$V_e = 0$ when both conditions occur:

$$V_e \geq \dfrac{V_u}{2}$$

And

$$P_u < \dfrac{A_g f_c'}{20}$$

Then, V_e = seismic shear demand, $V_c = 0$, and design stirrups to carry entire shear demand V_e within a distance 2H
  Special moment frame members subject to bending and axial loads ACI 21.6.1.
  Columns,

$$P_u > \dfrac{A_g f_c'}{10}$$

P_u includes earthquake loads
  The shortest cross-sectional dimension is 12"
  The ratio of short dimension to long dimension should not be less than 0.4, see Figure 7.17.
  Minimum flexural strength of column ACI 21.6.2

$$\Sigma M_{nc} \geq \left(\dfrac{6}{5}\right) \Sigma M_{nb} \qquad \text{ACI Equation 21.1}$$

where:
  ΣM_{nc} is sum of nominal flexural strengths of columns framing into the joint, evaluated at the faces of the joint
  ΣM_{nb} is sum of nominal flexural strengths of the beams framing into the joint, evaluated at the faces of the joint

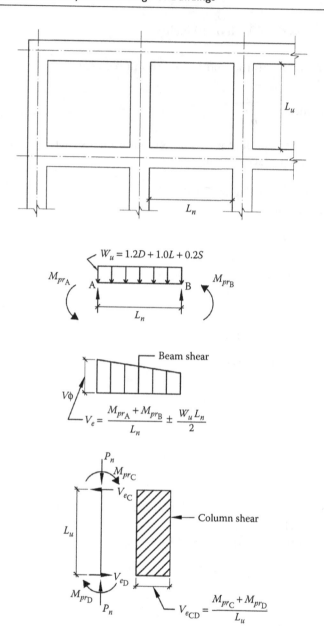

Figure 7.16 ACI requirements of shear in beams and columns in frames.

Longitudinal reinforcement ACI 21.6.3

$$0.01A_g \leq A_{st} \leq 0.06A_g$$

Transverse reinforcement ACI 21.6.4.

Hoops are required and a length l_0 from each joint face and shall not be less than, depth of member h, 1/6 of the clean span of member, and 18 inches

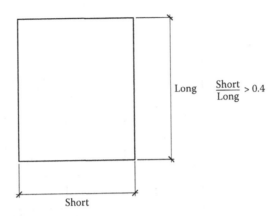

Figure 7.17 Rectangular column dimension ratio.

The total hoop area, A_{sh}, is the maximum as follows:

$$A_{sh} = 0.3 \frac{sb_c f_c'}{f_{yt}} \left[\left(\frac{A_g}{A_{ch}} \right) - 1 \right]$$ Equation 21.4

or

$$A_{sh} = 0.09 \frac{sb_c f_c'}{f_{yt}}$$ Equation 21.5

Spacing of transverse reinforcement ACI 21.6.4.3

Spacing of transverse reinforcement along the length l_0 of the member shall not exceed the smallest of the following:

1/4 of the minimum member dimension
Six times the diameter of the smallest longitudinal bar

$$s_0 = 4 + \left(\frac{14 + h_x}{3} \right)$$

And s_0 cannot exceed 6" or taken less than 4 inches, h_x = Spacing of cross-tie bars and shall not be greater than 14 inches

See Figure 7.18 for detailing of a special moment frame

Example 7.2: Special reinforced concrete moment frame joint design

This example demonstrates the procedure to determine the required joint relationship between beams and columns in a special reinforced concrete frame (Figure 7.19).
Flexural strength of columns must satisfy:

$$\Sigma M_{nc} \geq \left(\frac{6}{5} \right) \Sigma M_{nb}$$ ACI 21.62 Equation 21.1

$$M_{nb} = \frac{275^{k \cdot ft}}{0.9}$$

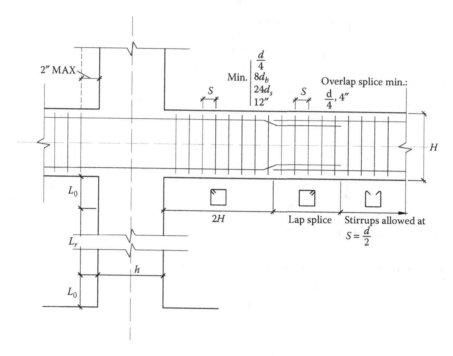

Figure 7.18 Required detailing of beams and columns in moment frames.

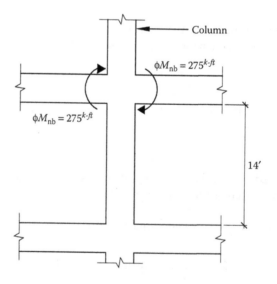

Figure 7.19 Beam moments at the face of joint of the frame.

$$M_{nb} = 306^{k \cdot ft}$$

$$\Sigma M_{nc} \geq \left(\frac{6}{5}\right) \Sigma M_{nb} \Rightarrow M_{nc} = \frac{1.2(306 + 306)}{2} = 367.6^{k \cdot ft}$$

Round up nominal moment capacity of column to $400^{k \cdot ft}$; this creates a strong-column and weak-beam scenario greater than the minimum requirement which is good design practice (Figures 7.20 and 7.21).

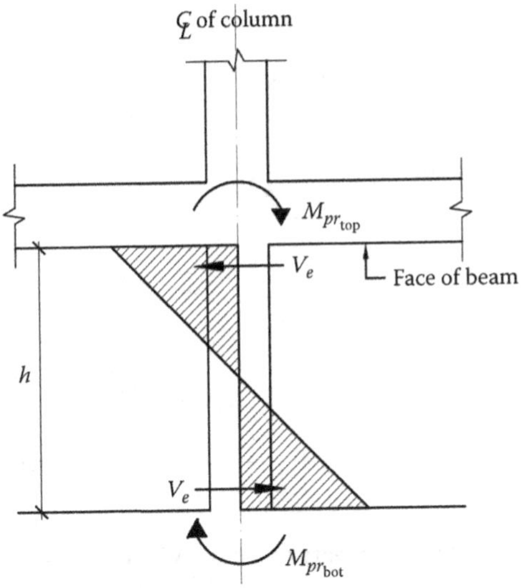

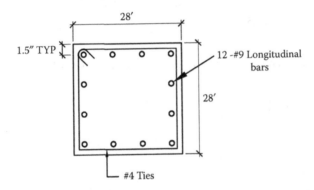

Figure 7.20 Column with moments and shears as expected with the frame action.

Figure 7.21 Cross section of the column.

The shear strength requirement ACI 21.5.4

V_e = the design shear force
M_{pr} = the probable flexural moment strength
V_e is based on M_{pr}

That is, $V_e = \dfrac{(M_{pr_{top}} + M_{pr_{bot}})1.25}{h}$

The moment can be designed in the column once the design is determined by proportioning the joint as required.

Moment capacity at beam-ends:

Use $1.25f_y$ and $\phi = 1.0$

$$V_e = \dfrac{\left(M_{pr_{top}} + M_{pr_{bot}}\right)1.25}{h}$$

$$= \dfrac{(400 + 400)1.25}{14} = 71.4\,\text{kips}$$

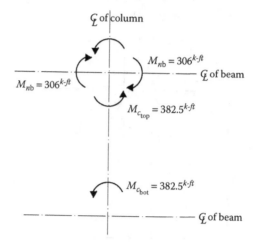

Figure 7.22 Calculated required column moments based on ACI seismic provisions.

V_e need not exceed shear based on girder M_{pr}. The beam nominal moment can be distributed to the column above and below the joint column moment below joint (Figure 7.22):

$$M_{c_{top}} = \frac{(306 + 306)1.25}{2} = 382.5^{k-ft}$$

The assumption is that the moment is transferred to the base as $M_{c_{bot}}$. Then,

$$V_e = \frac{\left(M_{c_{top}} + M_{c_{bot}}\right)}{h}$$

$$= \frac{(382.5 + 382.5)}{14} = 54.6\,\text{kips}$$

Thus, 54.6 kips controls for shear demand in the column.

Chapter 8

Wood

8.1 INTRODUCTION TO LATERAL WOOD DESIGN

8.1.1 Introduction and general information

The design standard for wood structures is the National Design Specification for Wood Construction (NDS) 2012. The International Building Code (IBC) is the governing general building code whereby Chapter 16 prescribes the required load combinations and references ASCE 7 for wind and seismic loading to be used for the design of wood structures. Chapter 23 of the IBC codifies the provisions for wood, which govern the materials, quality control, design, and construction of wood structures. The IBC references the NDS as the accepted standard for design accounting for the appropriate amendments and modifications noted in Chapter 23 as well as the use of the American Wood Council Special Design Provisions for Wind and Seismic (AWC SDPWS) 2008 for the design of lateral force-resisting systems.

The basic components of wood lateral systems are diaphragms (flexible), subdiaphragms, chords in diaphragms, diaphragm collectors, or drag struts, shear walls, shear wall chords, and shear wall hold downs. All these elements are design to work together and provide a later force resisting system.

Determination of design loads are as discussed in the previous chapters of this text but it is important to note that the source of loading is paramount to the design of the structure. That is, the source of the lateral force being due to wind or seismic determines the force load path, magnitude, and design parameters.

8.2 PLYWOOD DIAPHRAGM DESIGN

The diaphragm of wood structures is the major horizontal component, which transfers the lateral forces to the vertical shear wall elements. This transfer of force has an explicit load path and must be followed and traced accordingly to insure a proper functioning system.

Along with the analysis and design to resist forces, the building deformation is critical as discussed when determining serviceability and stability of the structure. As such deflection, determination is required to assess the building's potential movement.

Calculation of diaphragm deflection is based on bending, shear deflection, fastener deformation, chord splice slip and other contributing sources of deflection. The diaphragm deflection δ_{dia} shall be permitted to be calculated by the use of the following calculation:

$$\delta_{dia} = \frac{5vL^3}{8EAW} + \frac{0.25vl}{1000G_a} + \frac{\Sigma(x\Delta_c)}{2W} \qquad \text{SDPWS Equation 4.2-1}$$

where:

 E = modulus of elasticity of diaphragm chords, psi
 A = area of chord cross section
 G_a = Diaphragm shear stiffness (SDPWS Tables 4.2A, 4.2B, 4.2C or 4.2D) kips/in L = diaphragm length, ft
 W = diaphragm width, ft
 v = shear in the diaphragm, lbs/ft
 x = distance from the chord force to the nearest support
 Δ_c = diaphragm chord splice slip, inches
 δ_{dia} = mid-span diaphragm deflection, inches

Limiting aspect ratios (length to width) of diaphragms is established by AWC SDPWS and it governs the geometric design of diaphragm systems and subdiaphragm configuration. The subdiaphragms are a subordinate system to the larger system and should have explicit load paths to direct horizontal forces in the form of shear to the vertical load-resisting system of the building. Each subdiaphragm must have all the required components such as chords and struts and be fastened in accordance to sustain the shears they develop under loading. Limiting aspect ratios are a function of the material they are constructed from and the assembly of the system, that is, nailing pattern and blocked or unblocked. See Table 8.1 below.

Example 8.1: Plywood diaphragm analysis and chord splice design

This example illustrates an analysis of a flexible diaphragm and splice connector of a diaphragm chord.
  Wind pressure = 23 psf.
  The building shown in Figure 8.1 is a three-story wood-framed building. The building is 45 ft × 75 ft. The main level is 12 ft tall and the second and third levels are 10 ft tall. The wind load applied in the N–S direction is 23 pounds per square foot.

  1. Determine the design shear in pounds per liner foot for the second and third levels. Consider the floors and the roof to be flexible diaphragms. There are no internal shear walls (Figure 8.2).

     The design shear for the roof, the third- and the second-level diaphragms are calculated by multiplying the wind force by the associated tributary width as shown as follows (Figures 8.3 through 8.5):

Roof:

$$F_r = 23(5) = 115^{\#}/\text{linear foot}$$

Third level:

$$F_3 = 23(10) = 230^{\#}/\text{linear foot}$$

Table 8.1 Diaphragm aspect ratio

Maximum diaphragm aspect ratios (horizontal or sloped diaphragms)	
Diaphragm sheathing type	Maximum L/W ratio
Wood structural panel unblocked	3:1
Wood structural panel blocked	4:1
Single-layer straight lumber sheathing	2:1
Single-layer diagonal lumber sheathing	3:1
Double-layer diagonal lumber sheathing	4:1

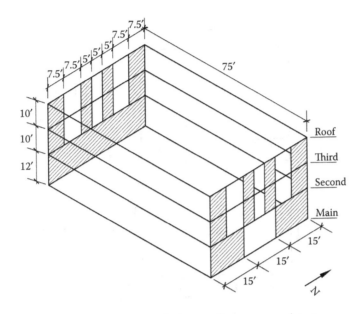

Figure 8.1 Three-story wood-framed building with shear walls systems as shown.

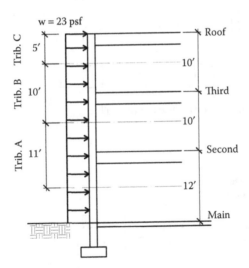

Figure 8.2 Building section showing wind pressures tributary to the floors and roof.

Second level:

$$F_2 = 23(11) = 253^\#/\text{linear foot}$$

To calculate the chord force, first calculate the maximum bending moment in the diaphragm. The chord force (tension or compression) is calculated by dividing the moment by the width of the diaphragm (d).

$$T_{chord} = C_{chord} = \frac{M}{d} = \frac{wl^2/8}{d} = \frac{wl^2}{8d} = \frac{253(75)^2}{8(45)} = 3953^\#$$

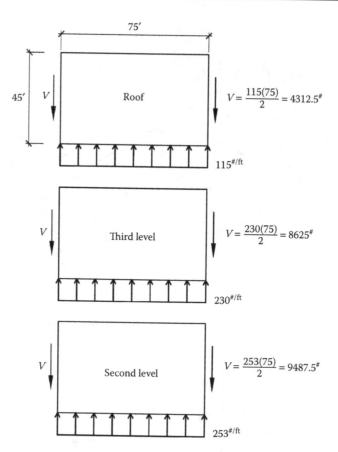

Figure 8.3 Diaphragm shear forces at each level of the building.

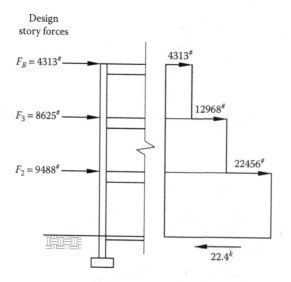

Figure 8.4 Design story forces and design story shears.

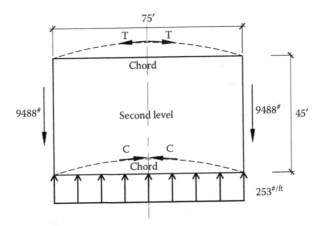

Figure 8.5 Diaphragm forces at the second level.

This building is a platform-framed structure, see Figure 8.6, and as such is built with a perimeter rim board, which if sized properly, can function as the chord and strut members of the diaphragm.

The 2 × 10 chord is 75 ft long on each side of the building and must be spliced together at several locations along its length. The splice connectors are typically constructed by end-butting two members together and overlapping another 2 × 12 member over the joint, see Figure 8.7. The connection is usually performed by bolting or nailing to fasten the splice on each side of the butt joint.

Select a 2 × 12 Douglas Fin-Larch (No. 1 and better) for the chord member. We will splice with the same material.

From the Natural Design Specifications (NDS) supplement, 2012 edition

Table 4A
Douglas Fin-Larch
No. 1 and better

$F_t = 800^{\#}$ (tension parallel to grain)

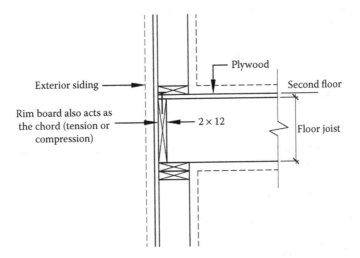

Figure 8.6 Section of building assembly through a typical floor.

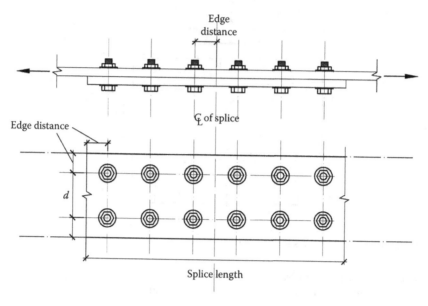

Figure 8.7 Bolted splice joint.

From 2012 NDS Table 2.3.2, the load duration factor $C_D = 1.6$ because the chord is subject to the wind load.

In addition, Table 4A of the NDS supplement 2012 notes additional adjustment factors that must be considered when using the values from this table.

They are as follows:

Size factor C_F applies to bending tension and compression parallel to grain design values for dimensional lumber 2″ to 4″ thick:

Grade: No.1 and better
    Width: 12″
    Thick: 2″
For the tension parallel to the grain F_t: $C_f = 1.0$

Repetitive member factor, C_r: does not apply
Wet surface factor, C_m: does not apply
Flat use factor, C_{fu}: does not apply
Incising factor, C_i: does not apply

NDS 2012 Table 4.3.1
The tension equation is as follows:

$$F_t' = F_t C_D C_m C_t C_f C_i$$

All adjustment factors are then equal to 1.0, with the exception of the duration factor, so the equation reduces to

$$F_t' = F_t C_D = 800(1.6) = 1280^{\#/in^2}$$

The actual tension parallel to the grains:

$$f_t = \frac{3594^{\#}}{1.5 \times 11.25} = 213^{\#/in^2}$$

$$f_t < F_t' = 1280^{\#/in^2}$$

Therefore, the 2 × 12 Doug Fir-Larch is more than adequate to sustain the tension force in the chord.

While the splice connector in Figure 8.7 illustrates bolts as the fasteners of the splice, because the tension force is not that large, nails may be a possibility. Therefore, try 10d common nails:

From 2012 NDS, Table 11N
Side member thickness = 1 ½ inches
Doug Fir-Larch
10d Common nails, Z = 118#

From NDS 2012 Table 10.3.1.
The connection equation is as follows:

$$Z' = ZC_DC_{d1}$$

$$C_{d1} = \text{diaphragm factor}$$

Section 11.5.3 NDS 2012

$$C_{d1} = 1.1$$

Therefore,

$$Z' = 118(1.6)1.1 = 207.7^{\#}$$

Then, calculate the number of nails per side of splice,

$$\text{No. of nails} = \frac{3953^{\#}}{208^{\#}/\text{nail}} = 19$$

Use 19 10d common nails per side of splice.

8.3 SHEAR WALLS AND COLLECTORS

Shear walls types are defined in AWC SDPWC Section 4.3. Shear walls are designed as segmented or perforated.

Segmented shear walls are full height wall segments where individual full-height wall segments are designed as shear walls without openings and like diaphragms have limiting aspect ratios. The limiting aspect ratios apply to each segment.

Wood structural panels, unblocked have an aspect ratio of 2:1, while blocked have 3.5:1.

The deflection of shear walls is extremely important; too large of a deflection of a diaphragm system can produce instabilities, where the components of the building do not continue to support loads as designed. Deflection limits are to be observed at all times.

Calculations of shear wall deflection shall account for bending and shear deflections, fastener deformation, anchorage slip, and other contributing sources of deflection.

The shear wall deflection, δ_{sw}, shall be permitted to be calculated by use of the following equation:

$$\delta_{sw} = \frac{8vh^3}{EAb} + \frac{vh}{1000G_a} + \frac{h\Delta_a}{b} \qquad \text{AWC SDPWS (4.3-1)}$$

where:

b = shear wall length, ft

Δ_a = total vertical elongation of wall anchorage system (including fastener slip, device elongation, rod elongation, etc.) at the induced unit shear in the shear wall, inches

E = modulus of elasticity of end posts, psi

A = area of end post cross section, in²

G_a = apparent shear wall shear stiffness from nail slip and panel shear deformation, kips/in. (from Column A, Tables 4.3A, 4.3B, 4.3C, or 4.3D)

h = shear wall height, ft

v = induced unit shear, lbs/ft

δ_{sw} = maximum shear wall deflection determined by elastic analysis, inches

Example 8.2: Diaphragm design and collector analysis

The second floor diaphragm of the building in Example 8.1 is subject to the seismic load, F_P, based on anchorage to flexible diaphragm force shown in Figure 8.8. Open web wood trusses span, N–S, 45 ft. They are spaced at 2 ft O.C., and have a 2″ nominal truss chord, which the sheathing is attached. Blocking is not required.

This example will explain the steps required to determine the appropriate sheathing thickness and nailing pattern for the diaphragm.

Use basic load combinations from IBC ASD:

$$U = 0.7E = 0.7(420) = 294^{\#/ft}$$

The shear at the east and west sides of the building is

$$V = \frac{294^{\#/ft}(75\ ft)}{2} = 11,025^{\#}$$

$$V = 11,025^{\#}/45\ ft\text{-long wall}$$

$$= 245^{\#/ft}$$

From AWC SDPWS 2008

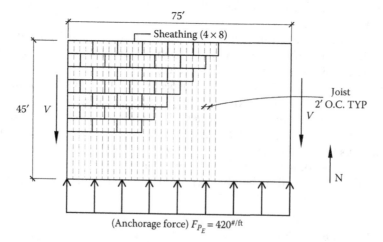

$$\text{(Anchorage force) } F_{P_E} = 420^{\#/ft}$$

Figure 8.8 Plan view of diaphragm with applied anchorage seismic design forces for flexible diaphragm.

Table 4.2C:
   Unblocked wood structural panel diaphragms
Case I sheathing pattern

Section: 4.2.3:
   ASD allowable unit shear capacities have a reduction factor of 2.0
   Enter Table 4.2C:
Seismic loading

Case 1
2″ Min. nominal width of nailed face (2″ truss chord)

$$V_s = \frac{510}{2.0 \text{ reduction factor}} = 255^{\#/ft} > 245^{\#/ft}, \text{ therefore OK}$$

Use sheathing 15/32
10d common nail size, 6 in nail spacing at diaphragm boundary, and supporting members
The main-level east and west elevations of the building are shown here with second the floor diaphragm unit shear forces shown (Figures 8.9 and 8.10)
The collector "gathers" the shear from the diaphragm and "pulls" the shear force into the wall or each side of the collector depending on the direction of the force. Collectors are also called drag struts

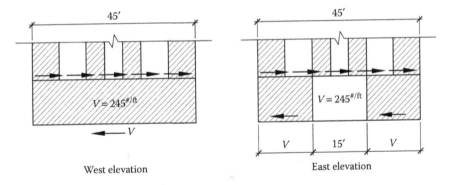

West elevation   East elevation

Figure 8.9 Partial east and west elevations (free-body diagrams).

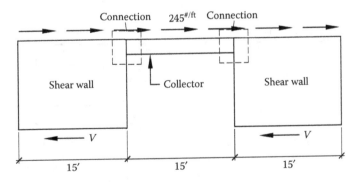

Figure 8.10 East elevation with collector between shear walls.

The collector's force is shared between the two adjacent walls
Hence, the force in the collector is

$$F_{collector} = \frac{245^{\#/ft}(15\ ft)}{2\ walls} = 1837.5^{\#}$$

Therefore, each connection between the collector and the wall has to be designed for this force
The shear at the base of the two 15 ft walls is

$$V_{base} = \frac{245^{\#/ft}(40\ ft)}{2\ walls} = 4900^{\#}$$

Example 8.3: Chord force determination in shear wall

Using the design story shears due to wind from Example 8.1, design the building's east side shear walls at the main level (Figure 8.11).

The design story shears act at the top of the main level wall produce an overturning moment. This moment is conservatively estimated to be resolved between the two ends of the wall (Figures 8.12 and 8.13).

From AWC SDPWS 2008, Table 4.3A.

Select sheathing and nailing for east wall pattern, based on the nominal unit shear capacities for wood frame shear walls.

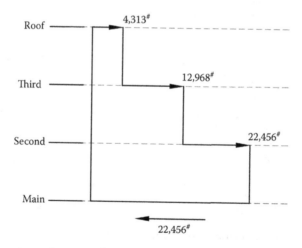

Figure 8.11 Design story shears due to wind.

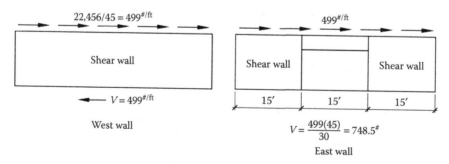

Figure 8.12 East and west shear walls with design story shears applied to walls.

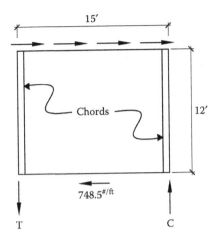

Figure 8.13 Discrete shear wall element of the east elevation.

Enter Table 4.3A:
Wind loading: B
Assume 2 × 4 stud framing

Then, based on a $V_w = \dfrac{1540^{\#/ft}}{2(\text{reduction factor Sec. 4.4.3)}} = 770^{\#/ft} > 748.5^{\#/ft}$ required, therefore OK

Use wood structural panels
Structural 1 – 15/32″
8d common nails
Panel edge fastener spacing = 3″

The west wall does not have as strong as a shear force as the east wall and it may be economical to size sheathing and nailing specific to its loads rather than use the same specifications for both sides.

Hence, enter the table again looking to satisfy $V = 770^{\#/ft}$ under wind loading:

$$V_w = \frac{1065^{\#/ft}}{2} = 532.5^{\#/ft} > 499^{\#/ft}, \text{ therefore OK}$$

Use 8d common nails, with 4″ edge.
   Fastener spacing, and
   15/32″ wood structural panels for sheathing.

Chapter 9

Masonry

9.1 INTRODUCTION TO LATERAL MASONRY DESIGN

The design standard for the design of masonry structures is the Building Code Requirements for Specifications for Masonry Structures (BCRSMS) 2013, which is known as TMS 402-13/ACI 530-13/ASCE 5-13. The International Building Code (IBC) is the governing general building code whereby Chapter 16 prescribes the required load combinations and references the use of ASCE 7 for wind and seismic loading to be used for the design of concrete structures. Chapter 21 of the IBC codifies the provisions for masonry, which govern the materials, quality control, design, and construction of concrete used in structures. The IBC references TMS 402-13/ACI 530-13/ASCE 5-13 as the accepted standard for design accounting for the appropriate amendments and modifications noted in Chapter 21.

As with all structural systems discussed, the lateral force-resisting systems detailing requirements are required to meet seismic detailing and limitations prescribed in IBC and ASCE 7 even when wind load effects are greater than seismic load effects. The building's structural design must meet the seismic requirements of Chapter 7 of BCRSMS based on the response modification factor even if the wind forces control the building design.

The design procedure for masonry structures is outlined here:

Step 1: Determine detail requirements for the building, that is, determine the SDC and determine the type of masonry system that will be required: ordinary, intermediate, or special design for the masonry elements.

Step 2: Start design with code required minimum reinforcement in chord elements and web or nonchord regions. Checking required minimum and maximum at the beginning of the design process allows this to be done.

Step 3: Determine grouting of wall and potential placement of reinforcement to anticipate weight of wall for design calculations.

The basic shear wall types are ordinary, intermediate, and special. The ductility of the system increases from an ordinary system design to an intermediate and to a special system design. The response modification factors reflect this as the R values increase with the more ductile system.

9.2 BUILDING WALL DESIGN FOR IN-PLANE LOADS

Example 9.1: Design of in-plane loaded wall

A single story building with masonry shear walls and a flexible roof diaphragm is shown in the figure below. This example will demonstrate (Figure 9.1) the following:

1. The design of a special reinforced masonry shear wall.
2. The out-of-plane moment and axial load at the mid-height of the wall and design of the wall.

Part 1

The building is in an SDC D site (Figure 9.2).
Seismic weight tributary to the roof level

$$w = 300,000^{\#}$$

Risk category II, $I_e = 1.0$
Mapped spectral response acceleration at short periods, $S_s = 1.40$
Mapped spectral response acceleration at 1-sec, $S_1 = 0.53$
Mapped long-period translation period, $T_L = 8$ sec
Redundancy factor $\rho = 1.0$

To determine the seismic shear forces on the walls, we will first calculate the base shear.
The site class is unknown or not determined. In this case, when the site class is unknown, site class D should be used for design.
ASCE 7, Section 11.4.2.
From ASCE 7, Table 11.4-1
Site class D

$$S_s = 1.4$$

Thus, $F_a = 1.0$

From ASCE 7, Table 11.4-2
Site class D

$$S_1 = 0.53$$

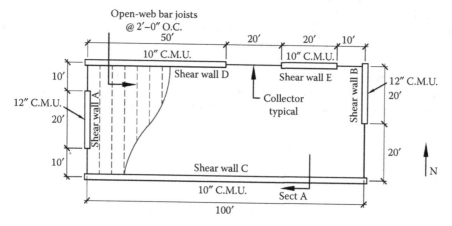

Figure 9.1 Plan view of masonry shear wall building with flexible roof diaphragm.

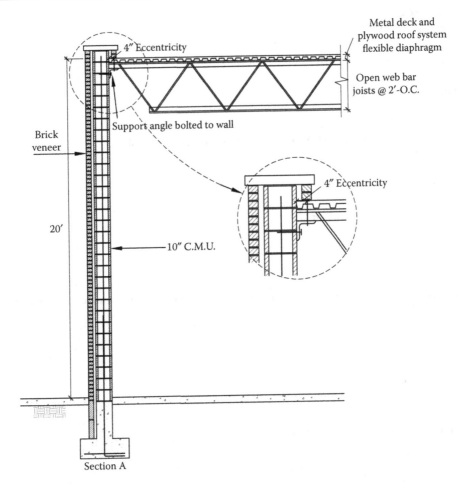

4" Eccentricity

Metal deck and
plywood roof system
flexible diaphragm

Open web bar
joists @ 2'-O.C.

Support angle bolted to wall

Brick
veneer

4" Eccentricity

20'

10" C.M.U.

Section A

Figure 9.2 Building section A.

Thus, $F_v = 1.5$

$S_{m_s} = F_a S_s = 1.0(1.4) = 1.4$

$S_{m_1} = F_v S_1 = 1.5(0.53) = 0.795$

$S_{D_s} = \dfrac{2}{3} S_{m_s} = \dfrac{2}{3}(1.4) = 1.0$

$S_{D_1} = \dfrac{2}{3} S_{m_1} = \dfrac{2}{3}(0.795) = 0.53$

From ASCE 7, Table 11.6-1
Risk category II

$S_{D_s} = 1.0 > 0.5$, then SDC = D

From ASCE 7, Table 11.6-2
   Risk category II

$S_{D_1} = 0.53 > 0.2$, then SDC = D

Therefore, SDC = D
   Select structural system
   From ASCE 7, Table 12.2-1
   A-7 special reinforced masonry shear wall (bearing wall system)
   No limit in height

$$R = 5, \Omega_0 = 2\frac{1}{2}, C_d = 3\frac{1}{2}$$

Determine seismic coefficient C_s

$$C_s = \frac{S_{DS}}{R/I_e} = \frac{1.0}{5/1.0} = 0.2$$

The period of the building can be calculated by using the approximate fundamental period
   ASCE 7 Section 12.8.2.1

$$T_a = C_t h_h{}^x$$

$$= 0.02(20)^{0.75} = 0.189$$

And C_s need not exceed the following:

$$C_s = \frac{S_{D_1}}{T(R/I_e)} = \frac{0.53}{0.189(5/1.0)} = 0.56$$

$$C_s = \frac{S_{D_1} T_L}{T^2(R/I_e)} = \frac{0.53(8)}{0.189^2(5/1.0)} = 23.74$$

And C_s must not be less than

$$C_s = 0.044 S_{D_s} I_e = 0.044(1.0) = 0.044 > 0.01$$

Use $C_s = 0.2$

The base shear is calculated by the following equation:

$$V = C_s w \hspace{4cm} \text{(ASCE and Equation 12.8-1)}$$

The seismic weight, $w = 300\,\text{kips}$
   Then,

$$V = 0.2(300) = 60\,\text{kips}$$

The seismic weight was given as part of this problem as 300 kips, and is typically the weight of the roof and the walls tributary to the roof. In this case, walls C, D, and E.

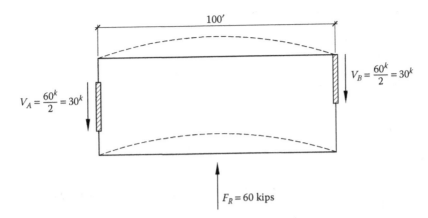

$V_A = \dfrac{60^k}{2} = 30^k$

$V_B = \dfrac{60^k}{2} = 30^k$

$F_R = 60$ kips

Figure 9.3 Building end shear based on flexible diaphragm behavior.

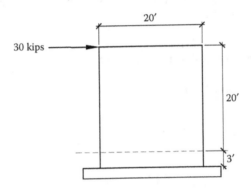

20′

30 kips

20′

3′

Figure 9.4 Lateral load applied to top of shear wall.

By applying the base shear in the North–South direction, we expect the roof diaphragm to distribute the seismic force evenly, based on tributary widths, to the end shear walls A and B (Figure 9.3).

$$F_x = C_{vx}V$$

$$C_{vx} = 1.0$$

$$F_R = 1.0V = 60^k$$

Seismic force applied to top of shear wall B (Figure 9.4).
    Shear wall design:
    Try 12″ solid grouted CMU wall

$$b = 11.625'', \ \gamma_{cmu} = 124 \ \text{psf}$$

$$f_m' = 2000 \ \text{psi}$$

$$F_s = 24,000 \ \text{psi} \quad \text{Grade 60 ksi}$$

Use alternate IBC load combinations,

1. $D+L+\left(L_r \text{ or } S \text{ or } R\right)$
2. $D+L+0.6wW$
3. $D+L+0.6wW+\dfrac{S}{2}$
4. $D+L+S+\dfrac{0.6wW}{2}$
5. $D+L+S+\dfrac{E}{1.4}$
6. $0.9D+\dfrac{E}{1.4}$

When shear walls, which have relatively small axial loads as compared to flexural loads and shear, that is,

$$M \gg P \text{ and } e > \frac{L}{2}$$

are considered tension design walls, which is a flexural design problem. Think of the wall as a vertical cantilevered beam.
Use ASD design:
Seismic load factor = 0.7 or E/1.4
Design wall for the loading shown in Figure 9.5.

$$P_{wall} = (23 \text{ ft})(20 \text{ ft})(0.124^{k/ft^2}) = 57.0 \text{ kips}$$

$$V_{wall} = \frac{C_s w_{wall}}{1.4} = \frac{0.2(57)}{1.4} = 8.14 \text{ kips}$$

$$\pm E_{v_{wall}} = \frac{0.2(S_{DS})D}{1.4} = \frac{0.2(1.0)57}{1.4} = \pm 8.14 \text{ kips}$$

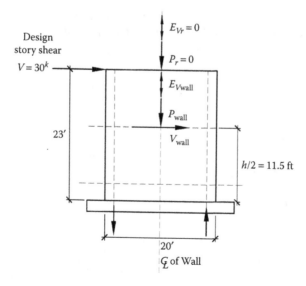

Figure 9.5 Free-body diagram of the wall.

Design loads (sum moments and forces at the base of the wall)

M_{max}:

$$M = V \cdot h + V_{wall} \cdot \frac{h}{2}$$

$$= 30(23) + (8.14)(11.5) = 783.6^{k \cdot ft}$$

Total shear:

$$V = V + V_{wall} = 30 + 8.14 = 38.14 \text{ kips}$$

Check C, T, and relationship: $e = M/P$

$$\text{Max } e = \frac{M}{0.9P_{wall} - E_{vwall}} = \frac{783.6^{k \cdot ft}}{[0.9(57) - 8.14]} = 18.14 \text{ ft} > \frac{L}{2}$$

Thus, $F_a \ll F_b$, flexural controls

$$\text{Max } F_a = \frac{P}{A} = \frac{P_{wall} + E_{vwall}}{A_{wall}} = \frac{(57 + 8.14) \times 1000}{11.625(20 \times 12''/ft)} = 23.34 \text{ psi}$$

$$\text{Min } F_a = \frac{P}{A} = \frac{P_{wall}}{A_{wall}} = \frac{57^k 1000^{\#/in}}{11.625(20 \times 12''/ft)} = 20.43 \text{ psi}$$

Create a trial flexural reinforcement configuration for analysis (Figure 9.6).

Examples of chord reinforcement configurations are as shown in Figure 9.7 and 9.8
Neglect nonchord reinforcement and use estimate of location for centroid of shear wall
Assume $j = 0.9$

$$d = L_{wall} - 16 \text{ in} = 20 \times 12 - 16 = 224 \text{ in} = 18.67'$$

$$A_s = \frac{M}{1.33F_s \cdot j \cdot d} = \frac{(783.6k \cdot ft)12''/ft}{1.33(24ksi)(0.9)(224)} = 1.46 \text{ in}^2$$

$$\text{No. of } \#8 \text{ bar} = \frac{1.46 \text{ in}^2}{0.44 \text{ in}^2} = 3.32 \text{ bars} \approx 4 \text{ bars}$$

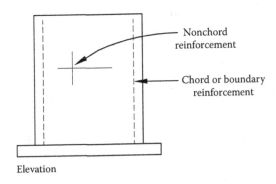

Nonchord reinforcement

Chord or boundary reinforcement

Elevation

Figure 9.6 Boundary reinforcement and nonchord reinforcement/web reinforcement.

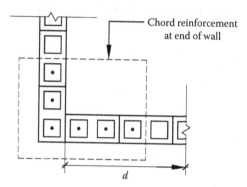

Figure 9.7 Reinforcement in walls with returns can be incorporated as boundary elements at ends of walls.

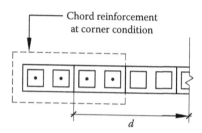

Figure 9.8 Straight wall boundary element.

Use 4-#8 bars (1 bar in last 4 cells)

$$A_s = 4(0.44) = 1.76 \text{ in}^2$$

Check maximum flexural compression

$$\rho = \frac{A_s}{bd} = \frac{1.76 \text{ in}^2}{11.625(224)} = 0.000676$$

$$E = 29{,}000 \text{ksi}$$

$$E_m = 900 f_m' = \frac{900(2000 \text{ psi})}{1000} = 1800 \text{ksi}$$

$$n = \frac{E}{E_m} = \frac{29000}{1800} = 16.11$$

$$\rho \cdot n = 0.000676(16.11) = 0.0109$$

$$k = \sqrt{(\rho n)^2 + 2(\rho n)} - \rho n = \sqrt{(0.0109)^2 + 2(0.0109)} - 0.0109 = 0.137$$

$$j = 1 - \frac{k}{3} = 1 - \frac{0.137}{3} = 0.954$$

$$f_b = \frac{2M}{j \cdot k \cdot b \cdot d^2} = \frac{2(783.6)12''/\text{ft}(1000 \, \#/k)}{(0.954)(0.137)(11.625)(224)^2} = 247 \text{ psi}$$

$$F_b = \frac{1}{3}f_m' = \frac{1}{3}(2000) = 666.7$$

Allowed to use 1/3 stress increase as per IBC 1605.3.2

$1.33 F_b > f_b$, therefore OK

Therefore, flexural compression is OK
   Check tension stress in steel

$$f_s = \frac{M}{A_s \cdot j \cdot d} = \frac{783.6(12''/\text{ft})}{1.76(0.954)(224)} = 25 \text{ ksi}$$

$f_s = 25 \text{ ksi} < 1.33 F_s = 1.33(24 \text{ ksi}) = 31.92 \text{ ksi}$, therefore OK

Check maximum compression in masonry
   Use load case $D+L+E$

$$r = \frac{d}{\sqrt{12}} = \frac{11.625}{\sqrt{12}} = 3.36$$

$$\frac{h}{r} = \frac{23 \times 12}{3.36} = 82.14 < 99 \qquad\qquad \text{BCRSMS Equation 8–16}$$

Then,

$$F_a = 0.25 f_m' \left[1 - \left(\frac{h}{140r} \right)^2 \right]$$

$$= 0.25(2000) \left[1 - \left(\frac{23 \times 12}{140(3.36)} \right)^2 \right]$$

$$= 327.9 \text{ psi}$$

The capacity is reduced by slenderness of the wall.

$D+L+E$

Max $F_a = 23.34 \text{ psi} < 327.9(1.33) = 435 \text{ psi}$, therefore OK

$D+L$

Min $F_a = 20.43 \text{ psi} < 327.9 \text{ psi}$, therefore OK

Quick compression check

$$\frac{f_a}{F_a} + \frac{f_b}{F_b} = \frac{23.34}{327.9} + \frac{247}{666.7} = 0.07 + 0.37 = 0.44 < 1.33, \text{ therefore OK}$$

Check shear
Special reinforced masonry shear walls having

$$\frac{M}{V_d} = \frac{783.6^{k \cdot ft}(12''/ft)}{30(224)} = 1.4 > 1.0$$

and

$$P > 0.05 f_m' A_n$$

then,

$$\rho_{max} = \frac{n f_m'}{2 f_y \left[n + f_y / f_m' \right]}$$

BCRSMS Equation 8.23

$$P = P_{wall} + E_{Vwall} = 57 + 8.14 = 65.14 \text{ kips}$$

$$\frac{0.05(7000)(24 \times 11.625)}{1000} = 260.4$$

Thus, $P < 0.05 f_m' A_n$

Then, do not have to check ρ_{max}

The calculated shear stress f_v shall not exceed the allowable shear stress F_v where F_v shall be calculated using BCRMS Equation 8.25:

$$F_v = (F_{vm} + F_{vs}) \gamma_g$$

where:

$$\frac{M}{Vd} > 1.0, \ F_{vmax} = \left(2 \sqrt{f_m'} \right) \gamma_g$$

BCRSMS Equation 8.27

$\gamma_g = 1.0$ for fully grouted

For special reinforced masonry, shear walls the actual shear stress is

$$F_v = \frac{1.5V}{bd}$$

and the allowable shear stress resisted by the masonry is

$$F_{vm} = \frac{1}{4} \left[\left(4.0 - 1.75 \left(\frac{M}{V_d} \right) \right) \sqrt{f_m'} \right] + 0.25 \frac{P}{A_n}$$

BCRSMS Equation 8.25

The allowable shear stress resisted by the steel reinforcement F_{vs} is calculated using

$$F_{vs} = 0.5 \left(\frac{A_v F_s d_v}{A_{nv} S} \right)$$

BCRSMS Equation 8.30

In other case,

$$\frac{M}{V_d} = 1.4 > 1.0$$

Therefore, the maximum, $F_v = 2\sqrt{f_m'} = 89.4$ psi

$$F_{vm} = \frac{1}{4}\left[\left(4.0 - 1.75\left(\frac{M}{V_d}\right)\right)\sqrt{f_m'}\right] + 0.25\frac{P}{A_n}$$

$$= \frac{1}{4}\left[(4.0 - 1.75(1.0))(44.72)\right] + 0.25\left[\frac{65.14 \times 1000}{(11.625 \times 224)}\right]$$

$\dfrac{M}{V_d}$ need not be taken greater than 1.0

$$= 78.26 + 6.25 = 84.51 < F_{vmax}, \text{ therefore OK}$$

Shear reinforcement shall be provided when F_v exceeds F_{vm}

$$F_v = \frac{1.5V}{bd} = \frac{1.5(30) \times 1000}{11.625(224)} = 17.28 \text{ psi}$$

$$F_v < F_{vm}$$

Therefore, additional reinforcement beyond minimum is not required for this wall (Figure 9.9).
    Calculate spacing of vertical and horizontal nonchord reinforcing.
    Horizontal reinforcing

$$A_g = 11.625 \times 20 \times 12 = 2790 \text{ in}^2$$

$$A_{shoriz} = 0.0007(2790) = 1.95 \text{ in}^2$$

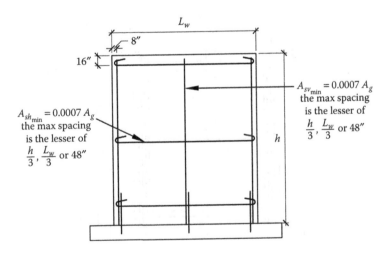

Figure 9.9 Minimum requirements of reinforcing for special masonry shear walls.

$$S_{max} = \frac{h}{3} = \frac{23}{3} = 7.6'$$

$$S_{max} = \frac{L_w}{3} = \frac{20}{3} = 6.67'$$

$$S_{max} = 48'' \qquad \rightarrow \qquad controls$$

$$A_{s_h} = \frac{1.95^2}{23 \text{ ft}} = 0.0847^{in^2/ft}$$

Using a #5 bar, $A_{\#5} = 0.31 \text{ in}^2$

$$S = \frac{0.31 \text{ in}^2}{0.0847^{in^2/ft}} = 3.66 \text{ ft} = 43.9 \approx 44''$$

Use #5 @ 44″ O.C. horizontally
Vertical reinforcing:

$$S_{max} = 48''$$

$$A_{s_h} = \frac{1.95^2}{23 \text{ ft}} = 0.0975^{in^2/ft}$$

Using a #5 bar

$$S = \frac{0.31 \text{ in}^2}{0.0975^{in^2/ft}} = 3.18 \text{ ft} = 38.15 \approx 38''$$

Use #5 @ 38″ O.C. vertically (Figure 9.10).

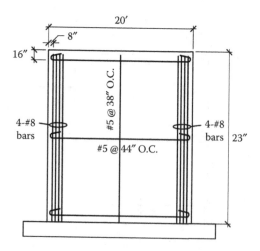

Figure 9.10 Final design of the special masonry shear wall.

9.3 BUILDING WALL DESIGN FOR OUT-OF-PLANE LOADS

Part 2. Axial and lateral load wall design

The out-of-plan seismic force is a function of the mass of the wall (Figure 9.11).

ASCE 7 Section 12.11 defines the out of plane force, which structural walls and their anchorages must be designed to sustain as $F_D = 0.4 S_{DS} I_e \times w_{wall}$.

The wall loading is both axial and flexural as shown in Figure 9.12.

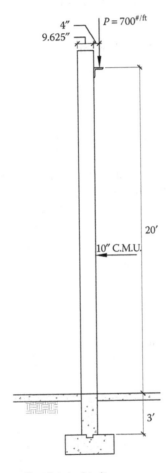

Figure 9.11 Free-body diagram of the wall with joist loading.

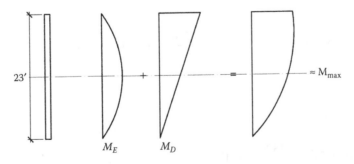

Figure 9.12 Diagram of seismic and dead load moments on wall.

Assuming a pin condition at the top and bottom of the wall.

The 10″ CMU walls are, assumed to be grouted, at 32″ O.C. The weight of the wall is $74^{\#/ft^2}$.

Then,

$$F_p = 0.4(1.0)(1.0)(74^{\#/ft^2}) = 29.6^{\#/ft^2}$$

The seismic out-of-plane force should not be less than 10% of the wall weight.

Then,

$$F_p = 29.674^{\#/ft^2} > 10\% \text{ of } 7474^{\#/ft^2}, \text{ therefore OK}$$

Design wall:

Assume pin–pin condition,

$$M_E = \frac{wl^2}{8} = \frac{29.6(23)^2}{8} = 1957.3^{\#\cdot ft}$$

The eccentricity due to the bearing point of the bar joist, which is offset from the face of the wall by 4″, is calculated from the center of the wall.

$$e = \frac{9.625}{2} + 4 = 8.81''$$

Then, the moment at the top of the wall is

$$M_{DL} = P \cdot e = 700^{\#/ft}\left(\frac{8.81 \text{ in}}{12''/ft}\right) = 513.9^{\#\cdot ft}$$

The moment at mid height of the wall is

$$M_{DL_{at\ mid-height}} = \frac{513.9}{2} = 257^{\#\cdot ft}$$

Use basic load combinations:

Moment:

5) $D + E/1.4$

$$(1.0 + 0.14S_{DS})D + \rho Q_E/1.4$$

$$\rho = 1.0$$

$$M_E = (1.0 + 0.14(1.0))(257) + \frac{1.0(1957.3)}{1.4} = 1691.05^{\#\cdot ft}$$

Axial load:

$$P_{Design} = P_{D_{roof}} + P_{D_{wall}}$$

$$= 700^{\#} + 74^{\#/ft^2}\left(\frac{23}{2}\right) = 1551^{\#}$$

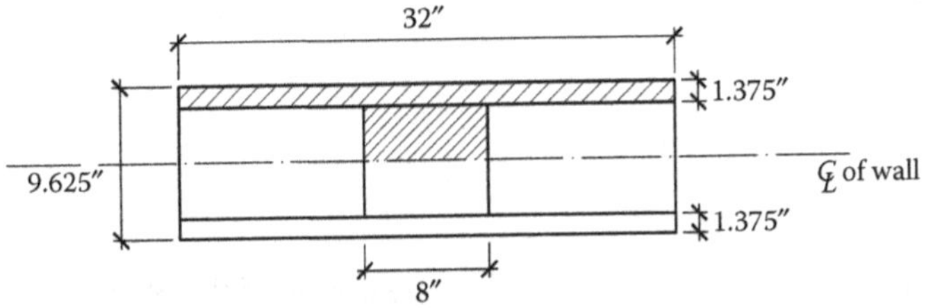

Figure 9.13 Cross section of wall which is assumed to contribute to resisting bending.

Partially grouted: assumed T-beam

$$M_{\text{Design}} = 1691.05^{\#\cdot\text{ft}}$$

$$P_{\text{Design}} = 1551^{\#}$$

Section properties of concrete masonry walls obtained from National Concrete Masonry Association, TEK 14 1A (Figure 9.13).
  Horizontal section properties
  10" single wythe walls @ 32" O.C.

$$A_n = 54.6^{\text{in}^2/\text{ft}}, \ I_n = 651.8 \ \text{in}^4$$

Axial stress:

$$f_a = \frac{P_{\text{mid-height}}}{A_n} = \frac{1551^{\#}}{54.6 \ \text{in}^2} = 28.4^{\#/\text{in}^2}$$

$$r = \sqrt{\frac{I_n}{A_n}} = \sqrt{\frac{651.8}{54.6}} = 3.45$$

$$\frac{h}{r} = \frac{23 \times 12}{3.45} = 80 < 99 \qquad\qquad \text{BCRSMS Section 8.2.4}$$

Thus, wall is not slender
  Then,

$$F_a = \frac{1}{4}f_m'\left[1 - \left(\frac{h}{140r}\right)^2\right] \qquad\qquad \text{BCRSMS Equation 8.16}$$

$$= \frac{1}{4}(2000)\left[1 - \left(\frac{80}{140}\right)^2\right] = 336.7^{\#/\text{in}^2}$$

$f_a < F_a$, therefore OK

Check stresses in masonry and steel:
   Steel stress:

$$F_s = \frac{M}{A_s jd}$$

$$d = \frac{9.625}{2} = 4.81 \quad \text{reinforcing is located in the center of the block}$$

$$E_m = 900 f_m' = 900(2000) = 1,800,000 \, \text{psi}$$

$$E_s = 29,000,000 \, \text{psi}$$

$$n = \frac{E_s}{E_m} = \frac{29}{1.8} = 16.11$$

Try #6 @ 32" O.C.:

$$A_{S\#6BAR} = 0.44 \, \text{in}^2$$

$$\rho = \frac{A_s}{b_w d} = \frac{0.44}{32(4.81)} = 0.00286$$

$$k = \sqrt{2\rho n + (\rho n)^2} - \rho n$$

$$\rho n = 0.00286(16.11) = 0.046$$

$$= \sqrt{2(0.046) + (0.046)^2} - 0.046$$

$$= 0.2607$$

$$kd = 0.2607(4.81) = 1.25'' < 1.375'', \text{ therefore OK}$$

The compression block is within the face shell (Figure 9.14).

$$j = 1 - \frac{k}{3}$$

$$= 1 - \frac{0.2607}{3} = 0.913$$

$$M_{Design} = 1691.05^{\#\cdot ft}/ft$$

$$M = 16901.05 \left(\frac{32''}{12''/ft} \right) = 4509.36^{\#\cdot ft} = 54,112.32^{\#\cdot in}$$

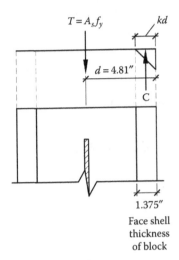

Figure 9.14 Section of concrete block with assumed compression block located within the face shell.

Steel stress:

$$f_s = \frac{54{,}112.32}{0.44(0.913)4.81} = 28{,}314.6^{\#/in^2}$$

$f_s = 28.3\,\text{ksi} < F_s = 1.33(24\,\text{ksi}) = 31.92\,\text{ksi}$, therefore OK

Check axial load capacity:

$$P_a = \left(0.25 f_m' A_n + 0.65 A_{st} F_s\right)\left[1 - \left(\frac{h}{140r}\right)^2\right]$$

$$= (0.25(2000)54.6 + 0.65(0.44)24{,}000)\left[1 - \left(\frac{80}{140}\right)^2\right]$$

$$= 23008^{\#} > 1551^{\#}, \text{ therefore OK}$$

Stress in masonry:

$$F_b = \frac{2M}{jkb_w d^2} = \frac{2(54{,}112.32)}{0.913(0.2607)(32)(4.81)^2} = 614^{\#/in^2}$$

$$F_b = \frac{1}{3} f_m' = \frac{1}{3}(2000) = 667^{\#/in^2}$$

Check combined axial compression and flexural stresses:

$$\frac{f_a}{F_a} + \frac{f_b}{F_b} \leq 1.33$$

$$\frac{28.4}{336.7} + \frac{614}{667} = 1.0 < 1.33, \text{ therefore OK}$$

Use #6 bar @ 32" O.C. for design of out-of-plane loading.

Chapter 10

Foundations and retaining structures

10.1 TYPES OF FOUNDATIONS

The foundation is the portion of the building or structure that transfers the building's loads to the soil or rock beneath the building. Foundations are designed to resist anticipated loads (which are based on the most current building code at the time the building is designed) that may occur during the life of the building, same as the other structural components of the building.

To assess physical properties, which influence the foundation selection and design parameters for a proposed building, the designer must understand the subsurface soil conditions. This requires specific testing and sampling of the soil and rock, if present, to determine allowable bearing capacities.

There are essentially three types of foundation systems commonly used in most building construction, depending upon existing site and soil conditions. They are spread footings, mat-slab foundations, and deep foundations.

Spread footings are typically designed for residential or moderate-height buildings with firm soil conditions present at a minimum. They are the most typical type of foundation used in building construction. Mat foundations are more complicated because they are a combined footing, which support both column and wall loads and redistribute the loads to the soil in a uniform bearing stress. Deep foundations are the most robust and versatile foundation system of the three, because of its adaptability to almost any site for almost any type of building structure. Deep foundations are designed to utilize the stiffer soil strata found deeper below the surface as well as bedrock if accessible. Deep foundations can also be designed to take advantage of friction developed on the surface of the foundation component in weak soils.

10.2 SPREAD FOOTING FOUNDATIONS

Spread footings are designed to "spread" or redistribute the load over the soil. To best describe this, let us consider a column that supports a vertical load of 100 kips, and it is to be supported by a spread footing that bears on a soil with a determined bearing capacity of 6.25 kips/ft². The required area needed to sustain the column load would be 100 kips/6.25 kips/ft², which equals 16 ft². Consequently, the footing's dimensions would be 4 ft × 4 ft. Thus, the column load is "spread" over 16 ft².

Spread footings are sized not only for bearing but also for overturning as well. That is, the footing's design is a function of the load being supported and the bearing capacity of the soil beneath the footing. Lateral loads will create an overturning moment on the footing, which will increase the pressure on the soil on one side of the footing and decrease the pressure on the other, as shown in Figure 10.1.

The bearing pressure of the soil determines the size of the bearing area of the footing needed to spread the load to the supporting soil. A spread footing is designed to resist the bending of the footing much like a cantilever beam would be designed and consequently is subject to beam shear, like a beam, as well.

For design purposes, the eccentricity, which is equal to the ratio of the overturning force to the vertical force, is usually kept to the length of the footing divided by 6.

$$\text{Eccentricity: } e = \frac{M}{P} \le \frac{L}{6}$$

The bearing pressure (q), as a function of the vertical force (P) and moment (M) applied to the footing is simply the force over the area of the footing (P/A) $\pm$ the applied moment divided by the section modulus of the plan area of the footing (M/S) (Figure 10.2).

$$\text{Bearing pressure: } q = \frac{P}{A} \pm \frac{M}{S} \tag{10.1}$$

We can rewrite $\dfrac{P}{A} \pm \dfrac{M}{S}$ as $\dfrac{P}{BL} \pm \dfrac{6Pe}{BL^2}$

Then, the bearing pressure or stress can be conventionally written as

$$q = \frac{P(1 \pm 6e/L)}{BL} \tag{10.2}$$

for a footing with an eccentricity located within the "kern" or $L/6$, of the footing.

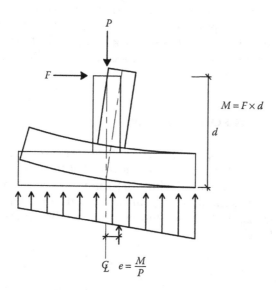

Figure 10.1 Axial and lateral load applied to a spread footing foundation.

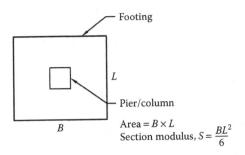

Footing

L

Pier/column

B

Area $= B \times L$

Section modulus, $S = \dfrac{BL^2}{6}$

Figure 10.2 Plan layout of square or rectangular footing.

10.2.1 Concentrically loaded footing

A concentrically loaded footing is simply a footing with the load applied at the center of the footing as in a column footing placed on an isolated spread footing or a wall load on a continuous spread footing.

A continuous wall footing or strip footing is primarily used to support exterior building walls and is designed for gravity loads as well as in-plane lateral loads. Exterior wall footings are subjected to frost in cold weather climates and should be placed at frost depth based on local building code requirements to prevent heaving.

The basic assumption for a concentrically loaded footing is as follows: The footing is considered to be *rigid* and the soil-bearing pressure is *uniform* and applied to the underside of the footing as shown in Figure 10.3. Following is a procedural approach for designing a concentrically loaded footing.

where:

q_a = allowable soil-bearing pressure

which is typically determined by soil testing or based on conservative assumptions based on visual observations.

q_e = effective soil-bearing pressure

which is determined by deducting the weight of the concrete and overburden soil on the footing, if any.

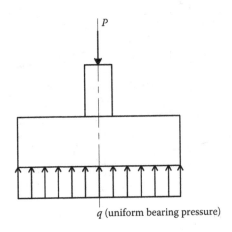

P

q (uniform bearing pressure)

Figure 10.3 Uniform bearing pressure due to column load.

And naturally the effective soil-bearing pressure must be less than the effective soil-bearing pressure: $q_e < q_a$.

Next is to determine the required bearing area using unfactored loads

$$\text{Footing bearing area required} = \frac{\text{Service load (unfactored loads)}}{q_e \text{ (effective bearing pressure)}}$$

Then, determine the footing thickness based on shear requirements, using factored loads.

1. Isolated spread footing
   a. Punching shear (b_o taken at $d/2$ from face of column)
   b. Beam shear (design section taken at d from face of column)
2. Continuous wall footing
   a. Beam shear (design section taken at d from face of column)

Then finally determine reinforcing based on the design bending moment at the face of the wall or column using factored loads.

Example 10.1: Concentrically loaded continuous wall footing

Determine the required width of the continuous wall footing, W, shown in Figure 10.4, when subjected to the following service (not factored) loads shown.

$w_{\text{dead load}} = 9.5 \text{ Kips/ft}$
$w_{\text{live load}} = 5.75 \text{ Kips/ft}$

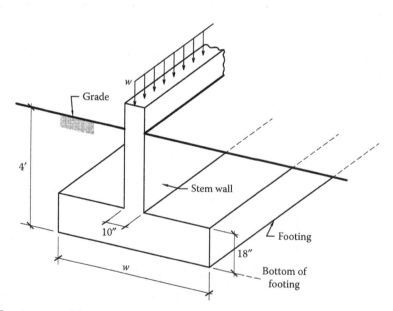

Figure 10.4 Continuous wall footing.

Given:

Compressive strength of concrete: $f'_c = 4000$ pounds/in^2 (psi).
Allowable soil-bearing pressure: $q_a = 1.75$ tons/ft^2 (tsf).
Soil density: $\gamma_s = 120$ pounds/ft^3 (pcf).
Concrete density: $\gamma_c = 150$ pounds/ft^3 (pcf).

Solution

Step 1: Determine the effective allowable soil pressure, q_e, by subtracting the weight of the footing and soil from the allowable soil pressure, q_a, on a per foot basis along the length of the footing.

Effective soil pressure, $q_e = q_a$–weight of concrete–weight of soil

Hence, $q_e = 1.75 \text{ tsf} \times 2000 \text{ lbs/ton} - \dfrac{18}{12}(150 \text{ psf}) - \dfrac{48-18}{12}(120 \text{ psf}) = 2975 \text{ psf}$

$$q_e = \frac{2975 \text{ psf}}{1000(\text{lbs/k})} = 2.98 \text{ ksf}$$

Step 2: The width of the footing can then be calculated by dividing the total service load by the effective soil pressure.

$$\text{Width, w} = \frac{\text{Service load}}{q_e} = \frac{w_D + w_L}{q_e} = \frac{9.5(\text{k/ft}) + 5.75(\text{k/ft})}{2.98 \text{ ksf}} = 5.1 \text{ ft, use w} = 5'0''$$

It is important to note that when calculating the effective bearing pressure and determining the size of the footing, always use nonfactored or service loads.

Once the width of the continuous wall footing is determined, the concrete footing itself must be checked for bending and beam shear.

Hence, to check the adequacy of the footing thickness, for the continuous wall footing in Example 10.3, the shear capacity needs to be determined.

The critical section for shear is located at distance, d, away from the stem of the wall as shown in Figure 10.5.

The effective bearing area creates a shear force across the failure or critical plane in the footing (Figure 10.6).

Solution

Step 1: Determine the strength design bearing pressure (q_u) using appropriate load factors:

$$q_u = \frac{w_u}{w(\text{length of wall})}$$

$$= \frac{1.2(9.5\text{k}) + 1.6(5.75\text{k})}{5'(1')}$$

$$= 4.12 \text{ ksf}$$

Step 2: Check beam shear along the critical section.

This is done by determining the ultimate shear (V_u) at the critical section, which is equal to the bearing pressure times the effective area, $V_u = q_u \times A_1$. A_1 is the effective bearing pressure area, shearing across the critical section (Figure 10.7).

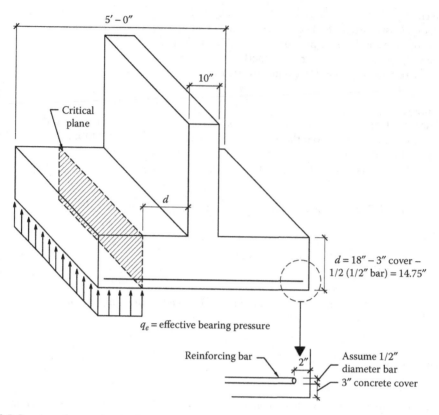

Figure 10.5 Portion of a continuous footing showing the failure plane for beam-shear due to the effective bearing pressure.

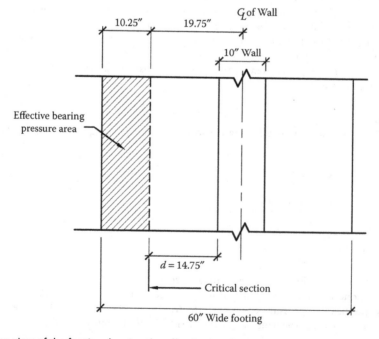

Figure 10.6 Plan view of the footing showing the effective bearing pressure area.

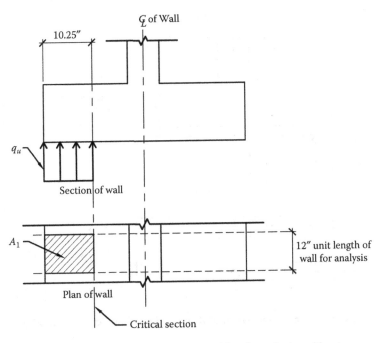

Figure 10.7 Plan and section of a 12″ unit width of wall used for shear design of footing.

For Example 10.1, the ultimate shear is then calculated as follows:

$$V_u = 4.12 \text{ ksf} \left(\frac{10.25″}{12″} \right) (1 \text{ ft unit width of wall}) = 3.52 \text{ Kips/linear foot}$$

Determine the shear strength of the footing and check adequacy to sustain the ultimate shear determined in the previous step.

The shear strength of the concrete must first be calculated as follows:

$$\phi V_c = \phi 2 \lambda \sqrt{f'c} \, bwd \qquad\qquad \text{(10.3) (ACI Equation 11.3)}$$

where:
ϕV_c is design strength of concrete
ϕ is found in section ACI 9.3.2.3 = .75
λ is normal weight of concrete found in section ACI 8.6.1 = 1.0
$f'c$ is compressive strength of concrete = 4000 psi
bw is unit length of wall = 12″
d is 14.5″

hence,

$$\phi V_c = .75(2)(1.0)\sqrt{4000 \text{ psi}} (12″)(14.5″) = 16,507 \text{ plf}$$

$$\phi V_c = 16,507 \text{ psf/1000 lbs/k} = 16.5 \text{ klf} > V_u = 3.52 \text{ klf. Therefore, OK.}$$

The footing thickness of 18 in. is more than adequate to sustain the ultimate shear produced by the bearing pressure acting on the bottom of the footing.

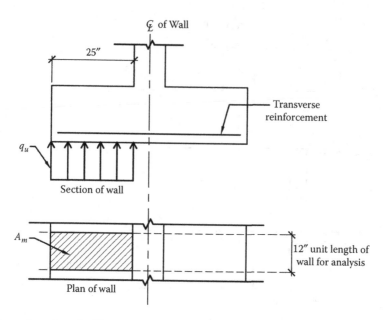

Figure 10.8 Plan and section of a 12" unit width of wall used for flexural design of footing.

Step 3: Determine the cross-sectional area of transverse reinforcing, A_s, required for flexure (Figure 10.8).

The design moment, M_u, is determined by assuming the flange of the footing to act as a cantilever beam extending from the face of the foundation wall. Therefore, the design moment can be calculated as follows:

$$M_u = \frac{wl^2}{2}, \quad w = q_u, \quad 1 = \frac{25 \text{ in}}{12 \text{ in/ft}} = 2.083 \text{ ft}$$

Hence, $M_u = (4.12 \text{ ksf})\dfrac{(2.083 \text{ ft})^2}{2} = 8.94 \text{ ft} - \text{kips/linear foot of footing}$

$$= 107.26 \text{ in-kips/ft}$$

Since the design moment is calculated per linear foot of footing, the area of transverse steel, A_s, should also be calculated for a 12-in. unit length of footing.

The actual steel ratio, ρ_{actual}, is readily calculated by the following equation:

$$\rho_{actual} = \frac{0.85f'c}{f_y}\left[1 - \sqrt{1 - \frac{2R_u}{0.85f'c}}\right], R_u = \frac{M_u}{\varphi bd^2} \tag{10.4}$$

where $\varphi = .90$, for flexure is found in section ACI 9.3.2.3

$$R_u = \frac{M_u}{\varphi bd^2} = \frac{107.26 \text{ in} - \text{kips}}{.9(12)(14.75)^2} = 0.0456 \text{ ksi}$$

$$= 45.6 \text{ psi}$$

Then,

$$\rho_{actual} = \frac{0.85(4,000)}{60,000}\left[1 - \sqrt{1 - \frac{2(45.6)}{0.85(4,000)}}\right] = 0.0008$$

$$\rho_{actual} < \rho_{minimum} = 0.0018$$

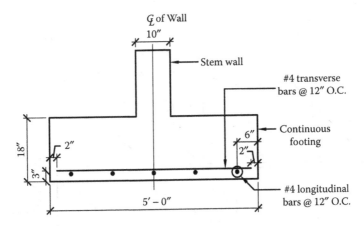

Figure 10.9 Completed continuous footing design.

The actual steel ratio is less than the minimum required to satisfy temperature and shrinkage requirements, $\rho_{minimum} = 0.0018$, as per ACI 318, 7.12

Therefore, the required $A_s = \rho_{min}(bh) = 0.018(12'')(18'') = 0.39$ in^2, where bh = gross concrete cross-sectional area.

For design, selecting a single #6 reinforcing bar at 12" O.C. yields $A_{\#6\ Bar} = 0.44$ in^2/ft, which is greater than the required $A_s = 0.39$ in^2 (Figure 10.9).

Hence,

use #6 at 12" O.C.

Example 10.2: Concentrically loaded isolated spread footing

Determine the required dimensions and flexural reinforcement required for the isolated square spread footing, subjected to the service loads shown, in Figure 10.10.

Given:

Compressive strength of concrete: $f'_c = 4000$ pounds/in^2 (psi)

Allowable soil-bearing pressure: $q_a = 2.0$ tons/ft^2 (tsf)

Soil density: $\gamma_s = 120$ pounds/ft^3 (pcf)

Concrete density: $\gamma_c = 145$ pounds/ft^3 (pcf)

Solution

Step 1: Determine the effective allowable soil pressure, q_e, by subtracting the weight of the footing and soil from the allowable soil pressure, q_a, on a per foot basis along the length of the footing.

Effective soil pressure, $q_e = q_a$–weight of concrete–weight of soil

Hence,

$$q_e = 2.0 \text{ tsf} \times 2000 \text{ lbs/ton} - \frac{18}{12}(145 \text{ psf}) - \frac{30}{12}(120 \text{ psf}) = 3483 \text{ psf}$$

$$q_e = \frac{3483 \text{ psf}}{1000(\text{lbs/k})} = 3.48 \text{ ksf}$$

Step 2: The required bearing area of the footing can then be calculated by dividing the total service load by the effective soil pressure (Figure 10.11):

$$\text{Required footing area} = \frac{\text{Service load}}{q_e} = \frac{P_D + P_L}{q_e} = \frac{175 \text{ k} + 125 \text{ k}}{3.48 \text{ ksf}} = 86.2 \text{ ft}^2$$

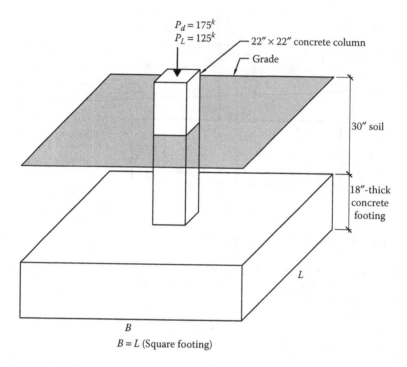

Figure 10.10 Concentrically loaded isolated spread footing; bottom of footing is located 48″ below grade.

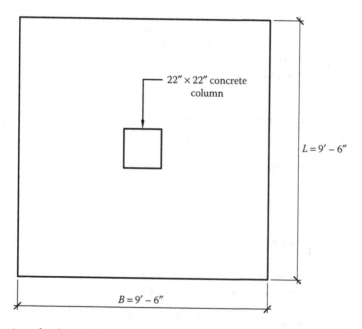

Figure 10.11 Plan view of isolated square spread footing, 9′–6″ × 9′–6″.

$B = L = \sqrt{86.2} = 9.3'$, round footing side up $9'-6''$

Therefore, $B = L = 9'-6''$.

Step 3: The footing thickness must be checked for adequacy against both *punching shear* and *beam shear*. This is done by first calculating the ultimate bearing capacity, q_u, using load factors, which will be applied to the underside of the footing (Figures 10.12 and 10.13).

$$q_u = \frac{P_u}{B \times L} = \frac{1.2(175)+1.6(125)}{9.5 \times 9.5} = 4.54 \text{ ksf}$$

Determine the design ultimate shear, V_u, generated by the ultimate loads acting on the footing, to be used to check the footing's punching shear adequacy.

$$V_u = q_u(A_p)$$

$$V_u = 4.54 \text{ ksf} \left[(9.5')(9.5') - \left(\frac{36.75''}{12''/\text{ft}} \right) \left(\frac{36.75''}{12''/\text{ft}} \right) \right]$$

$$V_u = 367 \text{ kips}$$

The punching shear capacity, φV_n, of the footing is found by the following formula:

$$\varphi V_n = \varphi V_c = \varphi 4\lambda\sqrt{f'c}b_o d \qquad \text{(10.5) (ACI Equation 11.33)}$$

The critical shear perimeter, $b_o = 36.75 \times 4 = 147''$.

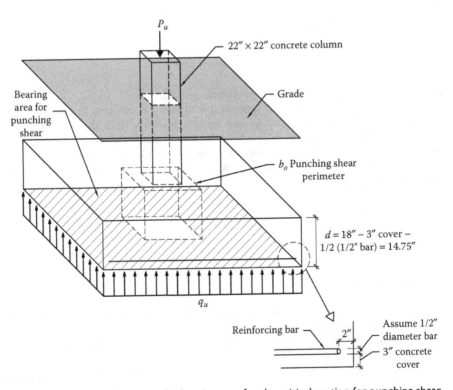

Figure 10.12 View of footing illustrating the bearing area for the critical section for punching shear.

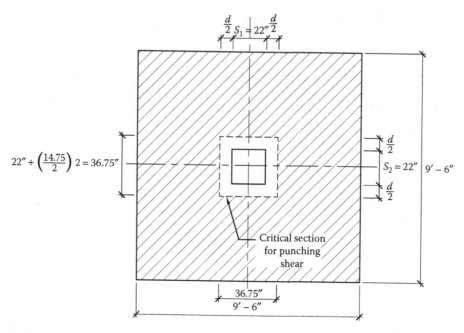

Figure 10.13 Plan view of footing illustrating the bearing area for the critical section for punching shear and the critical perimeter, b_o.

Hence,

$$\varphi V_n = 0.75(4)(1.0)\frac{\sqrt{4,000}}{1,000}(147)14.75''$$

$$\varphi V_n = 411 \text{ kips} > V_u$$

Therfore, the footing is OK (adequate) to sustain punching shear.

Next, determine V_u for beam shear or shear due to one-way bending. This is done similarly to the procedure used for a continuous footing; however, since the isolated footing has a finite width, in this case, 9'–6", we take the whole area as shown in Figure 10.14.

So, the shear force, V_u, across the critical plane, which is located at a distance d away from the face of the concrete column, is calculated by multiplying the bearing area by the ultimate bearing pressure.

$$V_u = q_u \times A_{pressure}$$

$$V_u = 4.54 \text{ ksf} \times \frac{31.25''}{12} \times 9.5 \text{ ft}$$

$$V_u = 112.3 \text{ kips}$$

Now, the shear strength of the footing must be checked for adequacy to sustain the ultimate shear by comparing it to the capacity of the footing's section.

The shear strength φV_n of the concrete, for one-way bending, is calculated as follows:

$$\varphi V_n = \varphi V_c = \varphi 2\lambda\sqrt{f'c}bwd \qquad\qquad (10.6) \text{ (ACI Equation 11.33)}$$

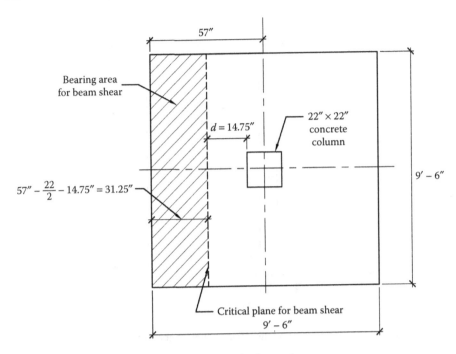

Figure 10.14 Plan view of footing showing bearing area for beam shear.

where:

φV_c is design strength of concrete
φ is found in section ACI 9.3.2.3 = .75
λ is normal weight of concrete found in section ACI 8.6.1 = 1.0
$f'c$ is compressive strength of concrete = 4000 psi
bw is length of wall = 9.5' × 12"/ft = 114"
d is 14.75"

hence,

$$\varphi V_c = .75(2)(1.0)\sqrt{4000 \text{ psi}}(114")(14.75") = 159,521 \text{ lbs}$$

$$\varphi V_c = 159,521 \text{ lbs}/1000 \text{ lbs/k} = 159.5 \text{ k} > V_u = 112.3 \text{ k}$$

Therefore, the depth of the section is OK.

The footing thickness of 18 in is more than adequate to sustain the ultimate shear produced by the bearing pressure acting on the bottom of the footing for both forms of shear.

Step 4: Now that we have the dimensions of our footing, we can now determine the cross-sectional area of transverse reinforcing, A_s, required for flexure (Figure 10.15).

The design moment, M_u, is determined by assuming the flange of the footing is acting as a cantilever beam extending from the face of the concrete column. Therefore, the design moment can be calculated as follows:

$$M_u = \frac{wl^2}{2}, \quad w = q_u, \quad 1 = \frac{46 \text{ in}}{12 \text{ in/ft}} = 3.83 \text{ ft}$$

Hence, $M_u = (4.54 \text{ ksf})(9.5 \text{ ft}) \dfrac{(3.83 \text{ ft})^2}{2} = 316.3 \text{ ft} - \text{kips}$

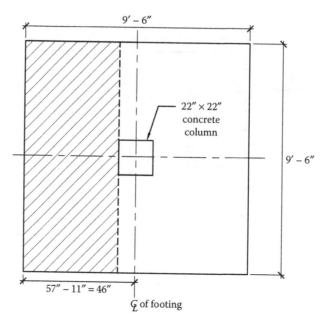

Figure 10.15 Bearing area used for calculating the design bending moment.

Since the design moment, here, is calculated for the full width of the isolated footing, the moment can be converted to a per linear foot of footing; then, the area of transverse steel, A_s, can be calculated for a 12-in unit length of footing.

$$M_u = \frac{316.3 \text{ ft} - \text{kips}}{9.5 \text{ ft}} = 33.3 \text{ ft} - \text{kips/foot of footing}$$

As with the continuous footing example, the actual steel ratio, ρ_{actual}, is readily calculated by the following equation:

$$\rho_{actual} = \frac{0.85f'c}{f_y}\left[1 - \sqrt{1 - \frac{2R_u}{0.85f'c}}\right], R_u = \frac{M_u}{\varphi b d^2} \qquad (10.4)$$

where $\varphi = .90$, for flexure is found in section ACI 9.3.2.3

$$R_u = \frac{M_u}{\varphi b d^2} = \frac{33.3 \text{ ft} - \text{kips}(12''/\text{ft})}{.9(12)(14.75)^2} = 0.1701 \text{ ksi}$$

$$= 170.1 \text{ psi}$$

Then,

$$\rho_{actual} = \frac{0.85(4000)}{60,000}\left[1 - \sqrt{1 - \frac{2(170.1)}{0.85(4000)}}\right] = 0.0029$$

$\rho_{actual} = 0.0029 > \rho_{minimum} = 0.0018$. Therefore, OK.

$$\rho_t = 0.319\beta_1\frac{f'c}{f_y} \qquad (10.7) \text{ (ACI Equation 11.33)}$$

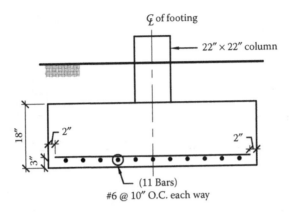

Figure 10.16 Completed footing design.

$\beta_1 = .85$ when $f_c' \le 4000$ psi

$$\rho_t = 0.319(0.85)\frac{4000}{60,000} = 0.0181$$

$\rho_{actual} = 0.0029 < \rho_t = 0.0181$. Therefore, OK.

Then, the required $A_s = \rho_{actual}(bd) = 0.0029(12'')(14.75'') = 0.51$ in²/ft width of footing. Selecting a single #6 reinforcing bar at 10" O.C. yields 0.53 in²/ft, which is greater than the required $A_s = 0.51$ in².

The spacing, s, and, $A_{s(actual)}$, calculations are as shown here.

Since

$A_{s(required)} = 0.51$ in²

and

$A_{\#6\ Bar} = 0.44$ in²

spacing of a #6 bar = $\dfrac{.44}{.51}(12\ \text{in}) = 10.35$ in, round down to 10" spacing.

$$A_{s(actual)} = \frac{.44}{10''\text{spacing}}(12\ \text{in}) = 0.53\ \text{in}^2/\text{ft}$$

Hence,
Use #6 @ 10" O.C. Each way (Figure 10.16).

10.2.2 Eccentrically loaded isolated spread footing

A footing designed to sustain an eccentricity due to an applied moment, created by a lateral force or an off-center gravity load, is considered an eccentrically loaded footing. Often, in design, the size of the footing is known due to a variety of constraints, such as proximity of adjacent footings and walls or property lines. So, starting an analysis and design of an eccentrically loaded footing is often begun with known dimensions and then the footing is modified accordingly.

To perform a footing design for an eccentrically loaded footing, the logical first step is to determine the soil reaction due to the eccentricity. That is, determine whether the soil reaction is bearing across the entire width of the footing, as with full bearing, or, bearing in part on the width of the footing, as with partial bearing.

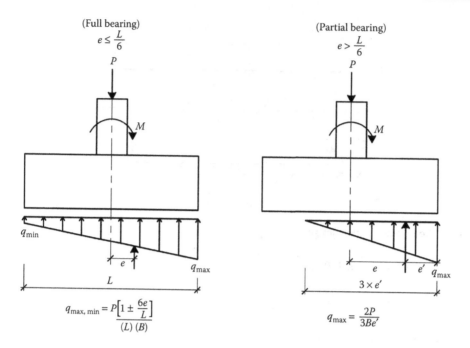

(Full bearing)

$$e \le \frac{L}{6}$$

q_{min} e L q_{max}

$$q_{max,\,min} = \frac{P\left[1 \pm \dfrac{6e}{L}\right]}{(L)\,(B)}$$

(Partial bearing)

$$e > \frac{L}{6}$$

e e' q_{max} $3 \times e'$

$$q_{max} = \frac{2P}{3Be'}$$

Figure 10.17 Full bearing and partial bearing soil reactions.

The soil reaction or bearing pressure distribution is a function of the eccentricity (e) on the footing, which is determined by dividing the moment (M) by the gravity load (P).

So, if $e = \dfrac{M}{P}$ is $\le \dfrac{L}{6}$, then the bearing pressure is considered to be full bearing or

when $e = \dfrac{M}{P}$ is $> \dfrac{L}{6}$, then the bearing pressure is partial bearing. See Figure 10.17 for bearing pressure equations.

After the soil pressure distribution is calculated, the footing design is carried out in much the same way as the previous footing design examples were performed. That is, calculate the factored shear and check the adequacy of the footing's thickness, and finally calculate the design moment and design the required flexural reinforcing.

Example 10.3: Eccentrically loaded isolated spread footing

The 24-in thick, 10 ft × 8 ft, isolated footing has a concrete column, which is built eccentric to the center of the footing, along the long dimension of the footing, as shown in Figure 10.18. The footing is supporting 100 kips of dead load and 90 kips of live load imparted by upper floors. The bearing capacity of the soil is as given below.

Determine if the bearing area of the footing is adequate for the allowable bearing pressure.

Given:

Compressive strength of concrete: $f'_c = 4000$ pounds/in² (psi)
Allowable soil-bearing pressure: $q_a = 3.0$ tons/ft² (tsf)
Concrete density: $\gamma_c = 145$ pounds/ft³ (pcf)

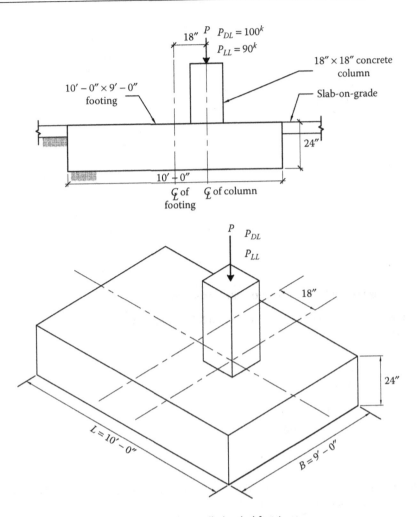

Figure 10.18 Dimensions and loading of an eccentrically loaded footing.

Typically, and as with this example, the footing design is begun with known dimensions, selected by the designer. So, the footing design becomes, effectively, a check to determine the adequacy of the bearing area and thickness of the footing to sustain the imposed loads.

Solution

Determine the soil pressure distribution along the width of the footing.

$$P_{column} = P_{DL} + P_{LL} = 100 \text{ kips} + 90 \text{ kips} = 190 \text{ kips (unfactored)}$$

$$P_{footing} = 10 \text{ ft} \times 9 \text{ ft} \times 2 \text{ ft} \times 145 \frac{\#}{ft^3} = 26.1 \text{ kips}$$

$$e = \frac{M}{P} = \frac{190(1.5)}{190 + 26.1} = 1.32'$$

$$\frac{L}{6} = \frac{10}{6} = 1.67 > e$$

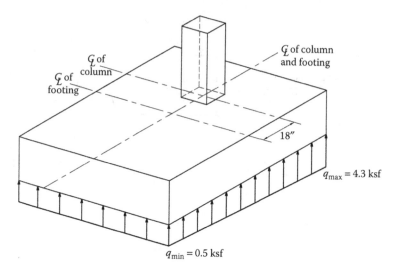

Figure 10.19 Bearing pressure due to eccentrically loaded footing.

Therefore, since the calculated eccentricity (e) is less than L/6, use the *full bearing* equation as shown in Figure 10.17.

$$q_{max,min} = \frac{P\left[1 \pm (6e/L)\right]}{(L)(B)} \tag{10.8}$$

$$q_{max,min} = \frac{(190 + 26.1)\left[1 \pm (6(1.32)/10)\right]}{(10)(9)}$$

$$q_{max} = 4.3 \text{ ksf}, \quad q_{min} = 0.5 \text{ ksf}$$

The effective bearing pressure is calculated by accounting for the weight of the footing
$q_e = q_a$-weight of concrete footing
$q_e = 3$ tons/sf (2k/ton)–2 ft (.145 k/ft³) = 5.71 ksf
Therefore,
$q_{max} = 4.3$ ksf $< q_e = 5.71$ ksf, therefore OK.
The maximum design bearing pressure is less than the effective bearing pressure (accounting for the weight of the footing) (Figure 10.19).

Example 10.4: Eccentrically loaded isolated spread footing

Consider the same footing in Example 10.3, with moment due to a wind load in the orthogonal direction as shown in Figure 10.20. Use the same soil-bearing capacity and determine if the footing size (bearing area) is adequate for the allowable soil-bearing pressure.

Solution
Use the following IBC load combinations (ASD) for dead load, live load, and wind load to determine the maximum soil-bearing capacity.

16–9	D + L
16–12	D + W
16–13	D + 0.75(W + L)
16–14	0.6D + W

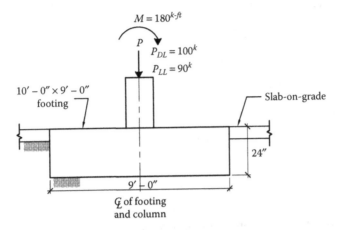

Figure 10.20 Orthogonal direction of footing in Example 10.3, with moment due to wind.

The allowable bearing capacity: $q_a = 3$ tons/ft^2
The concrete density: $\gamma_c = 145$ lb/ft^3
The weight of the footing: $P_{\text{footing}} = 10' \times 9' \times 2' \times 145 \text{ lb/ft}^3 = 26{,}000^{\text{lb}}$
Calculate the maximum and minimum bearing stress for each load combination:

1) $D + L$

$$q_{max} = \frac{P}{A} = \frac{P_{DL} + P_{\text{Footing}} + P_{LL}}{\text{Area of footing}} = \frac{100^k + 26.1^k + 90^k}{10' \times 9'} = 2.4^{\text{ksf}}$$

2) $D + W$
The wind load creates an eccentric loading on the footing, such that full or partial bearing must be determined.

$$e = \frac{M_w}{P_{DL}} = \frac{180^{k \cdot ft}}{100^k + 26.1^k} = 1.43'$$

$$\frac{L}{6} = \frac{9}{6} = 1.5 > e$$

Therefore, there is full bearing across the bottom of the footing; use Equation 10.6

$$q_{max,min} = \frac{P\left(1 \pm \dfrac{6e}{L}\right)}{(L)(B)} = \frac{126.1\left[1 \pm \dfrac{6(1.43')}{9'}\right]}{(10')(9')}$$

$$q_{max} = 2.74 \text{ ksf}, q_{min} = 0.07 \text{ ksf}$$

3) $D + 0.75(W + L)$

$$e = \frac{M_{0.75w}}{P_{D + 0.75L}} = \frac{0.75(180^{k \cdot ft})}{100^k + 26.1^k + 0.75(90^k)} = \frac{135^{k \cdot ft}}{193.6^k} = 0.7'$$

$$\frac{L}{6} = \frac{9}{6} = 1.5' > 0.7'$$

Therefore, full bearing

$$q_{max,min} = \frac{193.6\left[1 \pm \frac{6(0.7')}{9'}\right]}{(10')(9')}$$

$q_{max} = 3.16$ ksf, $q_{min} = 1.15$ ksf

4) $0.6D + W$

$$e = \frac{M_w}{P_{0.6D}} = \frac{180^{k \cdot ft}}{0.6(100^k + 26.1^k)} = 2.38'$$

$$\frac{L}{6} = \frac{9'}{6} = 1.5 < 2.38'$$

Therefore, partial bearing is present (Figure 10.21).
Use partial bearing equation as shown in Figure 10.17.

$$q_{max} = \frac{2P}{3Be'} \qquad (10.9)$$

$$q_{max} = \frac{2P_{0.6D}}{3(9')(2.12')} = 2.64 \text{ ksf}$$

The maximum soil pressure (q_{max}) is obtained from load case 3) $D + 0.75(W + L)$.

$q_{max} = 3.16$ ksf

The allowable bearing pressure was given as $q_{allowable} = 3$ tons/ft$^2 = 3 \times 2000 = 6000$ lb/ft^2
The effective soil-bearing pressure (q_e) is calculated as
$q_e = q_a$–weight of concrete footing
    $= 6000$ lb/ft$^2 - 156$ lb/ft$^3 \times 2'$-thick footing
    $= 5700$ lb/ft$^2 = 5.7$ ksf

$q_{max} = 3.16$ ksf $< q_e = 5.7$ ksf, therefore OK.

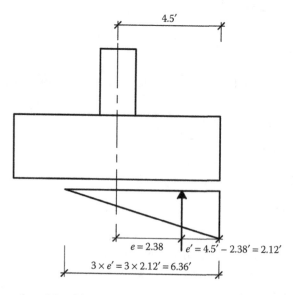

Figure 10.21 Distribution of partial soil-bearing pressure.

Example 10.5: Determine the factored design moment for an eccentrically loaded isolated spread footing

Consider the 10′ × 9′ footing discussed in both Examples 10.3 and 10.4. The soil-bearing pressure distributions for both directions are shown in Figure 10.22.

The applicable basic load combinations, as per the International Building Code (IBC), are as follows:

16–1	1.4D
16–2	1.2D + 1.6L
16–3	1.2D + 0.8W
16–4	1.2D + 1.6W + 1.0L
16–6	0.9D + 1.6W

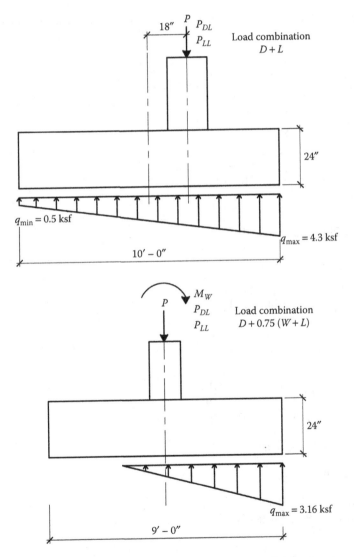

Figure 10.22 Long and short profiles of footing illustrating soil pressure distributions in each direction.

The governing load case in the long direction of the footing is load case 16-2:

$U = 1.2D + 1.6L$

$P_u = 1.2(100^k) + 1.6(90^k) = 264^k$

$$e = \frac{M}{P} = \frac{264^k(1.5')}{264^k + 1.2(26.1^k)} = 1.34'$$

$$\frac{L}{6} = \frac{10'}{6} = 1.67 > 1.34$$

$$q_{max,min} = \frac{P_u\left[1 \pm \dfrac{6e}{L}\right]}{(L)(B)} = \frac{264\left[1 \pm \dfrac{6(1.34')}{10'}\right]}{(10)(9)}$$

$q_{max} = 5.29$ ksf, $q_{min} = 0.57$ ksf

For the short direction of the footing, we will need to consider load cases 16-3, 16-4, and 16-6.

16-3: $U = 1.2D + 0.8W$

$$e = \frac{M_w}{P_{(1.2D)}} = \frac{0.8(180^{k \cdot ft})}{1.2(100^k + 26.1^k)} = 0.95'$$

$$\frac{L}{6} = \frac{9'}{6} = 1.5 > 0.95$$

$$q_{max,min} = \frac{P_u\left[1 \pm \dfrac{6e}{L}\right]}{(L)(B)} = \frac{151.3\left[1 \pm \dfrac{6(0.95')}{9'}\right]}{(10)(9)}$$

$q_{max} = 2.74$ ksf, $q_{min} = 0.62$ ksf

16-4: $U = 1.2D + 1.6W + L$

$$e = \frac{M_w}{P_{(1.2D+L)}} = \frac{1.6(180^{k \cdot ft})}{1.2(100^k + 26.1^k) + 90^k} = 1.2'$$

$$\frac{L}{6} > e$$

$$q_{max,min} = \frac{P_u\left[1 \pm \dfrac{6e}{L}\right]}{(L)(B)} = \frac{241\left[1 \pm \dfrac{6(1.2')}{9'}\right]}{(10)(9)}$$

$q_{max} = 4.82$ ksf, $q_{min} = 0.54$ ksf

16-6: $U = 0.9D + 1.6W$

$$e = \frac{M_w}{P_{(0.9D)}} = \frac{1.6(180^{k \cdot ft})}{0.9(100^k + 26.1^k)} = 2.54'$$

$$\frac{L}{6} < e$$

$$e' = \frac{9'}{2} - 2.54' = 1.96'$$

$$q_{max} = \frac{2P}{3Be'} = \frac{2[0.9(126.1)]}{3(10)(1.96)} = 3.86 \text{ ksf}$$

Consequently, load case 16-4 governs in the short direction of the footing ($q_{max} = 4.82$ ksf).

However, because the offset concrete column creates an eccentric loading, in the long direction, under dead and live load, and there is also a moment, due to wind, which, according to load combination 16-4, will occur simultaneously with the dead and live loads, the footing will experience bending in both directions at the same time, and thus, the footing is subject to biaxial bending.

Biaxial bending can be calculated using stress distribution (P/A) and (M/S) as follows:

$$q = \frac{P}{BL} \pm \frac{6Pe}{BL^2} \pm \frac{6M}{LB^2}$$

Using the factored loads from load case 16-4 (Figures 10.23 and 10.24):

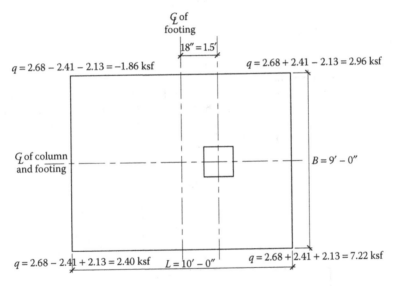

Figure 10.23 Plan view of footing illustrating the bearing capacity at each corner based on biaxial bending.

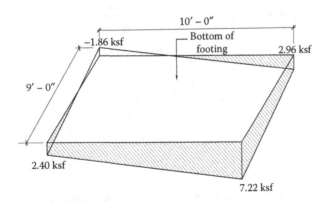

Figure 10.24 Resulting bearing capacities of footing at each corner.

$P_u = 241^k$

$M_u = 1.6(180) = 288^{k \cdot ft}$

then,

$$q = \frac{241}{(10)(9)} \pm \frac{6(241)1.5}{(9)(10)^2} \pm \frac{6(288)}{(10)(9)^2}$$

$$q = 2.68^{k/ft^2} \pm 2.41^{k/ft^2} \pm 2.13^{k/ft^2}$$

For design purposes, the resulting factored bearing stress, applied to the bottom of the footing, is conservatively estimated as shown in Figure 10.25.

The design moment can be calculated as follows for both sides of the column.

The moment at the left side of the column is

$$M_u = (2.40)(6.5)\left(\frac{6.5}{2}\right) + \frac{1}{2}(6.5)(3.13)\left(\frac{6.5}{3}\right) = 72.74^{k \cdot ft}$$

The moment at the right side of the column is

$$M_u = (2.40)(2)(1) + (3.86)(2)(1) + \frac{1}{2}(2)(0.96)\left(\frac{1}{3}\right) = 12.84^{k \cdot ft}$$

Clearly, the moment to the left of the column produces the critical or design moment for the footing to resist in bending.

The beam shear and punching shear are calculated and checked for adequacy in the same manner as discussed in Example 10.2. Albeit, the factored bearing stress distribution is more complicated due to the footing's eccentric loading, the procedure is exactly the same, and can be readily solved using a spread sheet in Excel.

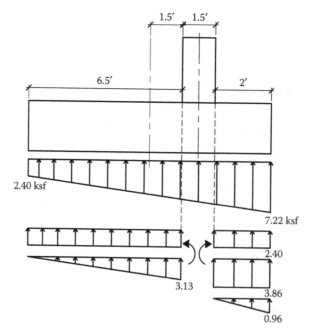

Figure 10.25 The bearing stress across width of footing used for determining design moment and shear.

10.3 MAT-SLAB FOUNDATIONS

A mat-slab foundation system, also referred to as a "raft," is essentially a large spread footing. Mat-slab or raft foundation systems are often used when poor subsurface soil conditions are present or when deep foundation systems are not feasible or less economical than a mat foundation solution. A mat foundation is designed to remain "rigid" in soft or weak soil strata, which offer very little in regard to bearing capacities. Mat-foundations are thought of as "floating" in soils of low bearing capacity because the mat or raft is designed to spread the wall and column loads across a large area of the foundation, which is in contact with the soil, and thereby reduce the required bearing capacity. Mat foundations must be sufficiently thick and capable of sustaining bending forces in order to ensure proper load transfer to the supporting soil layers (Figure 10.26).

Structural design of mat foundations is typically performed by two methods: the rigid method (conventional) or the approximate flexible method.

The rigid method is a direct application of the spread footing analysis and the flexible method is preferred when a mat foundation is to be located on expansive clays.

10.3.1 Combined footings

A combined footing is a type of mat foundation, which combines or accommodates wall and or column loads together in one footing, usually due to dimensional constraints or encroachments, which prevent the design of isolated spread footings (Figure 10.27).

> **Example 10.6: Combined mat footing design for two columns**
>
> A combined footing design is required for the column loads shown in Figure 10.28. Columns C1 and C2 are 18 ft apart and column C1 is 2 ft from the edge of the footing. Both columns are 18 in × 18 in. The thickness of the footing is assumed to be 24 in.

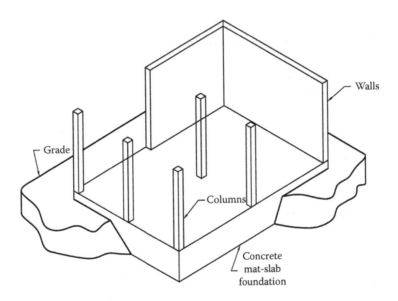

Figure 10.26 Mat-slab foundation system.

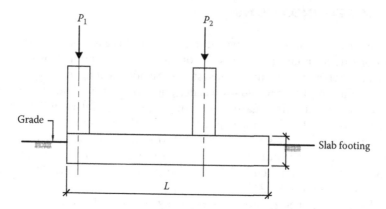

Figure 10.27 A combined footing or mat-slab foundation footing.

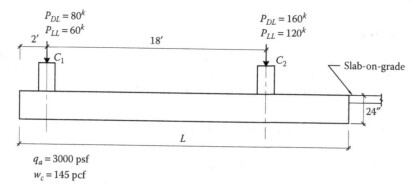

$q_a = 3000$ psf
$w_c = 145$ pcf

Figure 10.28 Combined footing supporting two columns.

Step 1: Locate centroid of loads by creating a free-body diagram and balance forces with the resultant (Figure 10.29):

$P_{C1} = 80 + 60 = 140^k$
$P_{C2} = 160 + 120 = 280^k$
$R = \Sigma P_i = P_{C1} + P_{C2} = 420^k$

$R\bar{x} = \Sigma P_i x_i = P_{C1}x_1 + P_{C2}x_2$

Solve for the centroid

$$\bar{x} = \frac{P_{C_1}x_1 + P_{C_2}x_2}{R}$$
$$= \frac{140^k(2') + 280^k(20')}{420^k}$$
$$= 14'$$

Step 2: Size footing such that the footing is centered about $\bar{x}$:
Length of footing $(L) = 2\bar{x} = 2(14') = 28'$

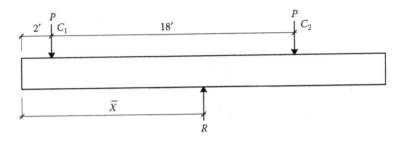

Figure 10.29 Free-body diagram of footing.

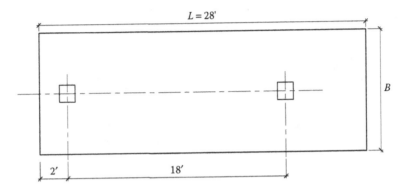

Figure 10.30 Plan view of footing.

Step 3: Determine the required width, B, of the footing:
The effective bearing capacity, q_e, is determined by subtracting the weight of the footing (Figure 10.30).

$$q_e = q_a - hw_c = 3000 \text{ lb/ft}^2 - 2'(145 \text{ lb/ft}^3) = 2710 \text{ lb/ft}^2 = 2.7 \text{ ksf}$$

Setting the effective bearing capacity equal to the total load, $P_{C1} + P_{C2}$, over the area of the footing, BL,

$$q_e = \frac{P_{C_1} + P_{C_2}}{BL}, \text{ rearrange terms, and solve for } B,$$

$$B = \frac{P_{C_1} + P_{C_2}}{q_e L} = \frac{140^k + 280^k}{2.7(\text{k/ft}^2)(28 \text{ ft})} = 5.56 \text{ ft}$$

Round up, the dimension, to the next foot and use 6′–0″ for the width of the footing.

We can now create the design shear and moment diagram for the combined footing as follows. The factored loads are as follows:

$$P_{uc_1} = 1.2(80) + 1.6(60) = 192 \text{ k}$$

$$P_{uc_2} = 1.2(160) + 1.6(120) = 384 \text{ k}$$

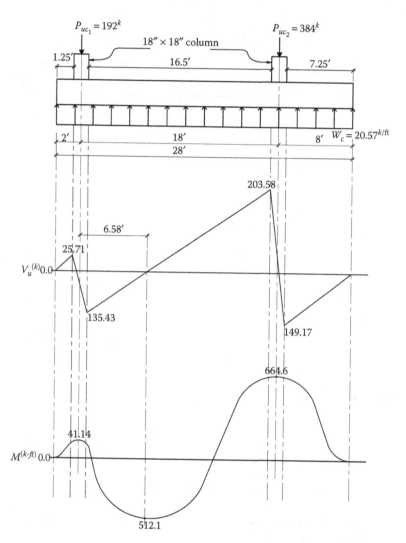

Figure 10.31 Design shear and moment diagrams.

and the uniform bearing capacity, w_u, is calculated as

$$w_u = \frac{P_{uc_1} + P_{uc_2}}{L} = \frac{192 \text{ k} + 384 \text{ k}}{28 \text{ ft}} = 20.57 \frac{\text{k}}{\text{ft}}$$

Using equations of statics, the shear and moment diagrams can be created as shown in Figure 10.31. For simplicity of design, the moments are taken, from the left, at 2', 9.33', and 20', which are the centerlines of the columns and the point of zero shear between the columns.

$$+ \wr M_{2'} = 20.57(2.0)(1) = 41.14^{\text{k·ft}}$$

$$+ \wr M_{9.33'} = 20.57(9.33)\frac{9.33}{2} - 192(7.33) = -512.1^{\text{k·ft}}$$

$$+ \wr M_{20'} = 20.57(20')\frac{20}{2} - 192(18) = 664.58^{\text{k·ft}}$$

10.4 DEEP FOUNDATIONS

Deep foundation systems utilizing pile caps and piles are reviewed in this section. The examples presented here are basic pile cap design procedures used in practice to quantify the adequacy of a design.

Deep foundation systems are engineered for buildings with large loads or in locations where the subsurface conditions are not adequate to sustain the imparted loads from the building. Sites with shallow levels of unsuitable soil strata will require a deep foundation system to reach suitable soil or rock with adequate bearing capacity at greater depths (Figure 10.32).

A pile cap is considered to be a reinforced concrete slab, which redistributes the imparted loads from a column or wall to a group of individual piles. A pile is a long slender member, which is commonly drilled, hammered (forced), or cast in place. The pile obtains its strength by end-bearing of the tip of the pile and by developing a surface frictional force known as skin friction or adhesion (Figure 10.33).

Example 10.7: Determine the maximum and minimum pile load

The maximum and minimum pile load is found by considering both the axial load distributed to the number of piles and the load imparted by the moment force as follows:

$$Q_{max/min} = \frac{P}{\text{number of piles}} \pm \frac{M}{S\,\text{pile group}} \tag{10.10}$$

$$S_{\text{pile group}} = \frac{I_{\text{pile group}}}{d_{\text{of pile group}}} \tag{10.11}$$

$$I_{\text{pile group}} = \Sigma n_i(d_i)^2 \tag{10.12}$$

So

$$I_y = 2(0) + 2(1.5')^2 + 4(3')^2 = 40.5 \text{ ft}^2$$

$\hookrightarrow$piles 2&7 $\hookrightarrow$piles 4&5 $\hookrightarrow$piles 1,3,6,8

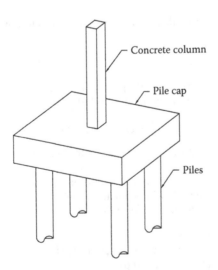

Concrete column

Pile cap

Piles

Figure 10.32 Structural elements of a deep foundation system.

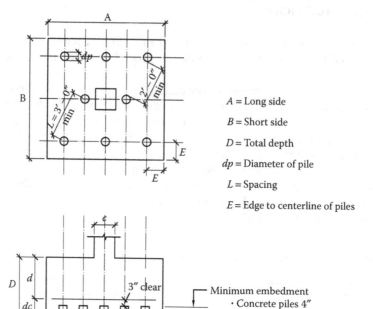

A = Long side

B = Short side

D = Total depth

dp = Diameter of pile

L = Spacing

E = Edge to centerline of piles

Minimum embedment
· Concrete piles 4″
· Steel piles 4″

E: 15″ minimum for pile capacity ≤ 60 tons
     21″ minimum for 60 < pile capacity ≤ 120 tons
     27″ minimum for 120 < pile capacity ≤ 200 tons
     30″ minimum for pile capacity > 200 tons

Figure 10.33 Pile cap nomenclature, dimensions, and plan and section, as per Concrete Reinforcing Steel Institute.

and

$$S_{\text{pile group}} = \frac{I_y}{d_{\max}} = \frac{40.5 \text{ ft}^2}{3 \text{ ft}} = 13.5 \text{ ft}$$

The maximum and minimum pile loading is then found as follows:

$$Q_{\text{max/min}} = \frac{570^k}{8 \text{ piles}} \pm \frac{800^{k\text{-ft}}}{13.5 \text{ ft}}$$

$Q_{\max} = 71.25^k + 59.26^k = 136.5^k$ (@ piles 3 and 8)
$Q_{\min} = 71.25^k - 59.26^k = 12^k$ (@ piles 1 and 6)

Example 10.8: Determine the required length of a pile

Determine the required length (L) of the maximum loaded pile in Figure 10.34, which was found to be 130.5^k.

In order to determine the length required, additional information of the proposed pile and the soil strata, where the pile will be located, is needed.

Consider the following additional information:

- 12″ diameter concrete piles
- Allowable end-bearing capacity, $q_{\text{tip}} = 20{,}000$ psf
- Allowable skin function, $f = 1200$ psf
- Factor of safety = 2

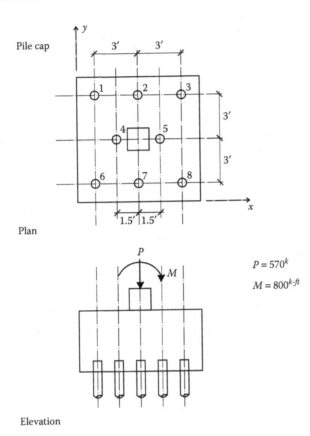

Plan

Elevation

$P = 570^k$
$M = 800^{k\text{-}ft}$

Figure 10.34 Plan and elevation of an 8-pile, pile cap subject to axial load and moment.

The pile capacity, based on the supporting strength of the soil, can be expressed in equation form as

$$Q_{\text{ultimate}} = Q_{\text{friction}} + Q_{\text{tip}}$$ (10.13)

where:

Q_{ultimate} is ultimate bearing capacity of a single pile
Q_{friction} is bearing capacity due to friction or adhesion between the soil and sides of the pile
Q_{tip} is bearing capacity due to soil just below the pile's tip

The term Q_{friction} in Equation 10.12 can be evaluated by multiplying the skin function (f) by the pile's surface (skin) area (A_{surface}). The term Q_{tip} is simply the area of the pile (A_{tip}) multiplied by the end-bearing capacity (q_{tip}).

Then, Equation 10.12 can be rewritten as follows:

$$Q_{\text{ultimate}} = f \cdot A_{\text{surface}} + q_{\text{tip}} \cdot A_{\text{tip}}$$ (10.14)

Hence, the solution is as follows:

$$A_{\text{tip}}, \text{ for a } 12'' \text{ diameter pile} = \frac{\pi d^2}{4} = \frac{\pi (1^{\text{ft}})^2}{4} = 0.785 \text{ ft}^2$$

$$A_{\text{surface}} = 2\pi r(L) = 2(\pi)(0.5 \text{ ft}) = 3.14 \text{ ft}(L)$$

Then, using Equation 10.13, we can write

$130.5^k (2) = 1.2 \text{ k/ft}^2 (3.14 \text{ ft}) L + 20 \text{ k/ft}^2 (0.785 \text{ ft}^2)$

Note, we must include the factor of safety by multiplying the pile load by 2. Then, solving for L,

$$L = \frac{130^k (2) - 20^{k/ft^2}(0.785^{ft^2})}{1.2^{k/ft^2}(3.14^{ft})} = 64.8' \approx 65'$$

The pile length must include the embedment length in the pile cap. Consequently, for a concrete pile, the minimum embedment length is 6"; therefore, the design length of pile must be $L = 65' + 0.5' = 65.5'$.

It must be noted here that the determined length of the pile, based on subsurface soil conditions, is obviously not the structural strength of the pile. A pile's structural strength is determined by the size and shape and material composition of the pile. The pile must be designed specifically for its design loads. Selection of materials for the pile, such as timber, reinforced concrete, and steel will require different design methods, which are not covered here but can be found in various design aids based on prevailing building codes.

Example 10.9: Determine the design shear and moment in a pile cap

Consider the pile cap shown in Figure 10.35 and determine the shear and moment at section A-A, on the pile cap (Figure 10.36).

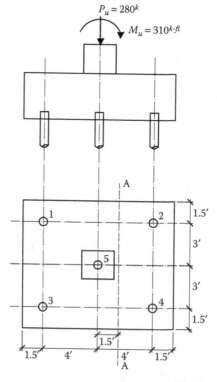

Figure 10.35 Plan and elevation of a 5-pile, pile cap subject to axial load and moment.

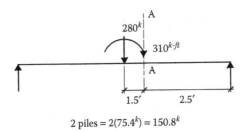

$$2 \text{ piles} = 2(75.4^k) = 150.8^k$$

Figure 10.36 Free-body diagram of pile cap.

From Equation 10.9,

$$Q_{max/min} = \frac{P}{n} \pm \frac{M_{yx}}{\Sigma x^2} \pm \frac{M_{xy}}{\Sigma y^2}$$

where:
  $P = 280^k$
  $n = 5$ piles
  $M_y = 310^{k \cdot ft}$
  $\Sigma x^2 = 4$ piles $(4 \text{ ft})^2 = 64 \text{ ft}^2$

Then,

$$Q_{@\,pile2} = \frac{280^k}{5} + \frac{310^{k \cdot ft}(4^{ft})}{64^{ft^2}} = 75.4^k$$

$$\curvearrowright M_u = 150.8^k(2.5') = 377^{k \cdot ft} \text{ (taken at section A-A)}$$

$$V_u = 150.8^k$$

Example 10.10: Pile loads due to shear wall loading

Given the wall and pile cap configuration in Figure 10.37, determine the loading on piles due to the shear wall loads. Note that the shear wall is asymmetrical to the pile cap and the piles are not evenly spaced (Table 10.1).
  The solution type of problem can be completed in three steps:

Step 1. Find the center of gravity of the pile group
Step 2. Determine the design moment about the center of gravity
Step 3. Determine the load on each of the piles using

$$Q = \frac{P}{n} \pm \frac{\Sigma m d_i}{\Sigma d_i^2}$$

Step 1.
  Locating the center of gravity is best done in a tabulated form as follows:

$$\text{Area of pile} = \frac{\pi(1^{ft})^2}{4} = 0.785 \text{ ft}^2$$

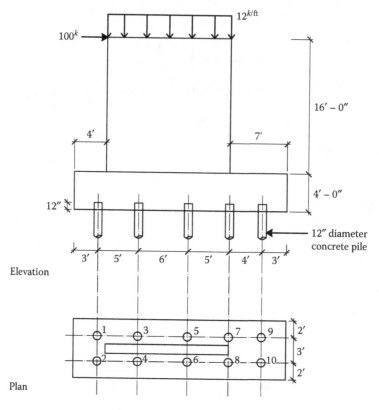

Figure 10.37 Plan and elevation of an axially and laterally loaded shear wall supported by a 10-pile, pile cap.

Table 10.1 Location of center of gravity of pile group

Pile	Area (ft²)	X₍ᵢ₎ (ft)	AᵢXᵢ (ft³)
1,2	2(0.785) = 1.57	3	4.71
3,4	1.57	8	12.56
5,6	1.57	14	21.98
7,8	1.57	19	29.83
9,10	1.57	23	36.11
	7.85		105.19

$$\bar{x} = \frac{\Sigma A_i x_i}{\Sigma A_i} = \frac{105.19}{7.85} = 13.4 \text{ ft}$$

Hence, the center of gravity is 13.4 ft from the left most edge of the pile cap.

Step 2.

Set up a free-body diagram to depict the loading on the pile cap about the center of gravity and sum moments (Figure 10.38).
A clockwise moment about the c.g. is shown here.

$$+\, \circlearrowleft\, M_{c.g} = 100^k \left(19^{ft}\right) - 12^{k/ft}\left(15\right)\left[13.4 - 11.5\right] = 1558^{k \cdot ft}$$

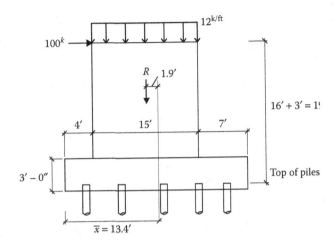

Figure 10.38 Free-body diagram of wall loading on pile cap.

The equivalent point load (R) of the uniform load is $12^{k/ft}(15') = 180^k$. This load is located at the center of the wall and to the left of the c.g.

Step 3.
Quantifying the loads on each pile is done in tabulated form using the familiar equation as shown (Table 10.2).

$$Q = \frac{P}{n} \pm \frac{\Sigma m d_i}{\Sigma d_i^2}$$

In summary, the minimum and maximum pile loads are at piles 1 and 2, which experience an uplift force of 13.02 kips, and piles 9 and 10, which experience a downward force of 46.63 kips. It is important to note that the weight of the footing was not accounted for in this problem and must be ultimately incorporated for a properly designed system.

10.5 RETAINING STRUCTURES

The design of retaining structures is based on the determination of lateral earth pressure and surcharges being imposed on the structure. Retaining structures, which are designed to be either braced or freestanding, see Figure 10.39, will promote a specific type of lateral

Table 10.2 Loads on each pile

Pile	d_i (ft)	d_i^2 (ft²)	$\frac{P}{n}$ (k)	$\frac{\Sigma m d_i}{\Sigma d_i^2}$	$P = \frac{P}{n} \pm \frac{\Sigma m d_i}{\Sigma d_i^2}$
1,2	13.4 − 3 = 10.4	2(108.16)	$\frac{180}{10} = 18$	$\frac{1558(10.4)}{522.4} = 31.02$	= 18 − 31.02 = −13.02 uplift
3,4	13.4 − 8 = 5.4	2(29.16)	18	16.1	= 18 − 16.1 = 1.9 uplift
5,6	14 − 13.4 = 0.6	2(0.36)	18	1.79	= 18 + 1.79 = 19.79 uplift
7,8	19 − 13.4 = 5.6	2(31.36)	18	16.7	= 18 + 16.7 = 34.7
9,10	23 − 13.4 = 9.6	2(92.16)	18	28.63	= 18 + 28.63 = 46.63
		522.4			

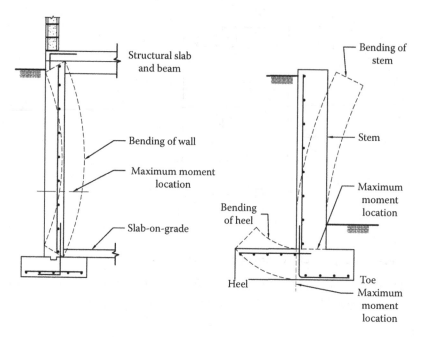

Figure 10.39 Illustration of bending of a subsurface foundation wall and a free-standing cantilevered retaining wall.

soil pressures. These pressures are considered to be static, active, or passive. Analysis of the retaining structure to resist the lateral soil pressures and surcharges is necessary from a stability and member design perspective.

Because the type of soil pressure present is in part based on the proposed structure, that is, a braced-wall system will induce a static soil pressure and a free-standing system will promote active and passive lateral soil pressures, careful attention to the selection of the structural system and associated parameters is required for proper analysis, design, and application.

10.5.1 Foundation walls

A foundation wall, such as a basement wall, which is braced at the floor and slab levels, is essentially an unyielding wall with virtually no lateral movement. A wall, which does not move, or at least very little, consequently develops a lateral soil pressure, on the wall, which is considered to be static earth pressure. The static earth pressure is obtained by considering the unit weight or density of the soil the retaining structure is located within, and if the soil is submerged in water or not, and multiplying it by the coefficient of static lateral earth pressure K_o.

The coefficient of static lateral earth pressure (K_o) ranges from 0.4 for dense sand to 0.5 for loose sand and can be calculated by the following formula:

$$K_o = 1 - \sin\Phi \tag{10.15}$$

where Φ is the angle of internal friction of the soil, determined by physical testing of the soil.

The following example will utilize static lateral soil pressure to analyze a foundation wall.

Example 10.11: Determine the maximum bending moment and location on basement foundation wall

Determine the maximum bending moment and location on the basement foundation wall shown in Figure 10.40.

The density (γ_s) of the backfill soil against the wall is $100^{\#/ft^3}$, and the angle of internal friction (Φ) of the soil is 36°.

A basement wall will bend the same as a simply supported beam would bend, if the top and bottom of the wall is sufficiently braced by the floor level at the top of the wall and the slab on grade at the bottom of the wall. The floor joists, because they are perpendicular to the wall, act as struts within the floor diaphragm. When the floor joists are parallel to the wall, bridging between the joists may be added to create adequate bracing in order to consider the wall braced. This is an important detail and should be conveyed in the full design of the wall and floor systems as part of the design documents.

Considering the wall to be braced at the top and bottom, we can use a static lateral earth pressure for design. The coefficient of lateral earth pressure can be calculated as follows:

From Equation 10.14,

$$K_o = 1 - \sin \Phi$$
$$K_o = 1 - \sin (37°) = .398$$

The soil pressure is determined by multiplying K_o by the unit weight of the soil γ_s.

$$\text{Unit lateral earth pressure} = (.398)\left(100\frac{\#}{ft^3}\right) = 39.8\frac{\#}{ft^2}, \text{ use } 40\frac{\#}{ft^2}$$

Hence, the pressure at the bottom of the wall is a function of the height of the soil, bearing against the wall, see Figure 10.41.

The reaction at A can be determined by summing moment about B:

$$\circlearrowleft + M_B = 0 = \frac{1}{2}(10')\left(400\frac{\#}{ft^2}\right)\left[\frac{10'}{3}\right] - A(10')$$

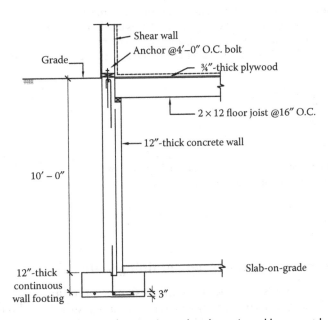

Figure 10.40 Section through basement foundation braced at the main and basement levels.

Labels in figure: Shear wall; Anchor @4′–0″ O.C. bolt; Grade; ¾″-thick plywood; 2 × 12 floor joist @16″ O.C.; 12″-thick concrete wall; 10′ – 0″; 12″-thick continuous wall footing; Slab-on-grade; 3″

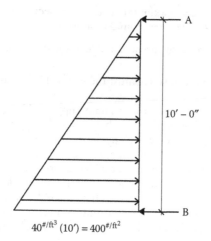

$$40^{\#/ft^3} (10') = 400^{\#/ft^2}$$

Figure 10.41 Free-body diagram of wall with static lateral earth pressure distribution.

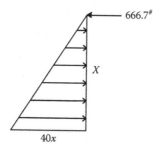

Figure 10.42 A free-body diagram to locate zero-shear along the wall.

Solving for A yields a reaction of 666.7#.

Because the maximum moment occurs where there is zero shear, to find the location of the moment, we need to locate zero shear on the wall; this can be done as follows (Figure 10.42):

Setting the reaction at A equal to an equation which represents the soil pressure loading of the wall as a function of height, the exact location where the equation is zero can be found.

$$666.7^{\#} = \frac{1}{2}(x)(40x)$$

$$33.3 = x^2$$

$$5.77 = x$$

Then, sum up moments at 5.77′ from the top of the wall, which will yield the maximum moment on the wall (Figure 10.43).

$$\curvearrowleft + M_{5.77} = 0 - \frac{1}{2}(5.77')(40 \times 5.77)\left[\frac{5.77'}{3}\right] + 666.7(5.77') = 2566^{\#-ft}$$

10.5.2 Free-standing cantilevered retaining walls

The lateral earth pressure used for analysis and design of a retaining wall structure is commonly determined using Rankine theory of lateral earth pressures. Rankine theory is based on soil, which has no adhesion or friction between the wall and soil, that is, the wall is

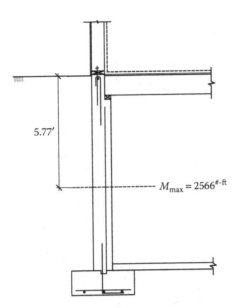

Figure 10.43 Maximum bending moment and location from top of the basement wall.

smooth. The soil being retained (active soil) and the soil in front of the structure (passive soil) is assumed to slide or fail along an incline or failure plane defined as a function of the soil's angle of internal friction (Φ). Hence, the failure of the soil wedge, along the failure plane, in the case, when the wall is moved by the soil, is said to be active pressure and when the wall moves the soil, in front of the structure, is said to be passive pressure, as discussed in Chapter 1.

For the majority of situations where the backfill surface, adjacent to the retaing wall is level, the coefficients for active and passive lateral pressures, using Rankine theory, are calculated by the following equations:

$$K_a = \frac{1 - \sin\Phi}{1 + \sin\Phi} \tag{10.16}$$

$$K_p = \frac{1 + \sin\Phi}{1 - \sin\Phi} \tag{10.17}$$

The retaining wall shown in Figure 10.44 has a surcharge at the top of the wall and the resulting pressures from the soil, surcharge, and the bearing pressure distribution at the bottom of the footing are described for use in quantifying design loads.

Pressures are calculated as follows:

$P_a = \gamma_s K_a$—active soil pressure
$P_p = \gamma_s K_p$—passive soil pressure
$P_L = W_L K_a$—live load soil pressure

The friction force created between the base of the footing and the soil due to the weight of structure is simply the product of the weight of the structure and the coefficient of friction (μ).

$$F = \mu\left(W_s + W_w + W_b\right)$$

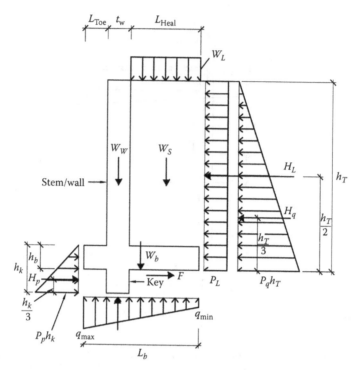

Figure 10.44 Illustration of loading of a cantilevered retaining wall.

The steps for analysis can be as listed here.

Step 1: Determine lateral earth pressures of soil based on Rankine theory
Step 2: Calculate unfactored sliding forces and overturning moment at the toe
Step 3: Proportion footing using industry factor of safety (FS), such that
    a. $M_{righting} > M_{overturning}$ (FS), where $1.5 < FS < 2.0$
    b. $H_{resist} \geq H_{slide}$ (FS), where FS > 1.5
Step 4: Use factored loads to design stem or wall and footing, based on ACI principals for walls and footings

Example 10.12:   Determine the adequacy of a cantilevered retaining wall

The 20-ft high retaining wall, shown in Figure 10.45, has a 300 psf surcharge at the top of the wall and has 1 ft of soil over the toe of the footing. Determine the adequacy of the wall by calculating the factor of safety for sliding and overturning, as well as calculating the bearing pressure at the toe and heel of the footing. Additional information is as follows:

Density of the soil, $\gamma_s = 120 \dfrac{\#}{ft^2}$

Weight of concrete, $\gamma_c = 150 \dfrac{\#}{ft^2}$

Angle of internal friction, $\emptyset = 35°$
Coefficient of friction, $\mu = 0.4$

It is typical to draw the retaining wall with the pressure distributions of the loads acting on the wall as shown in Figure 10.46, and locate the resultants of the lateral pressures.

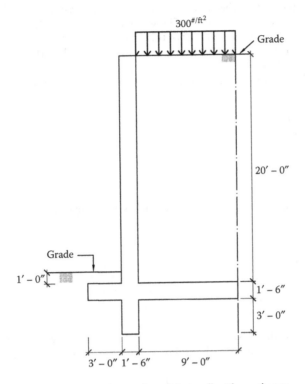

Figure 10.45 A 20-ft high free-standing cantilevered retaining wall with surcharge.

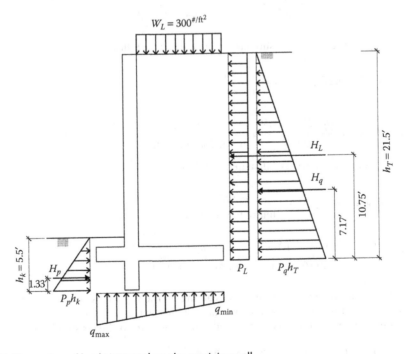

Figure 10.46 Illustration of loads imposed on the retaining wall.

It is important to note that the passive pressure is being considered for the full depth of the footing below the grade at the toe. The soil above the toes is often conservatively not considered due to the possible removal of this soil during the life of the structure.

To determine the lateral earth pressure, the Rankine lateral earth pressure coefficients are calculated as follows.

From Equation 10.15, the coefficient of active lateral earth pressure is calculated as follows:

$$K_a = \frac{1-\sin\varnothing}{1+\sin\varnothing}$$

$$K_a = \frac{1-\sin35}{1+\sin35} = 0.271$$

and similarly, from Equation 10.16, the coefficient of passive lateral earth pressure is as follows:

$$K_p = \frac{1+\sin\varnothing}{1-\sin\varnothing}$$

$$K_p = \frac{1+\sin35}{1-\sin35} = 3.69$$

Then, the lateral earth pressures can be calculated on a, per foot, linear length of wall. The active pressure is calculated as follows:

$$P_a = K_a\gamma_s = 0.271\left(120^{\#/ft^3}\right) = \frac{32.52^{\#/ft^2}}{\text{linear foot of wall}}$$

$$H_a = \frac{P_a b_T^2}{2} = \frac{\left(32.52^{\#/ft^2}\right)(21.5')^2}{2\left(1000^{\#/ft}\right)} = \frac{7.52^k}{\text{linear foot of wall}}$$

The lateral pressure due to the surcharge is calculated as follows:

$$P_L = K_a W_L = 0.271(300^{\#/ft^2}) = \frac{81.3^{\#/ft^2}}{\text{linear foot of wall}}$$

$$H_L = P_L\left(b_T\right) = \frac{\left(81.3^{\#/ft^2}\right)21.5'}{1000^{\#/ft}} = \frac{1.75^k}{\text{linear foot of wall}}$$

Passive pressure (resistance) is calculated as follows:

$$P_p = K_p\gamma_s = 3.69(120^{\#/ft^3}) = \frac{442.8^{\#/ft^2}}{\text{linear foot of wall}}$$

$$H_p = \frac{P_p b_k^2}{2} = \frac{\left(442.8^{\#/ft^2}\right)(5.5')^2}{2\left(1000^{\#/ft}\right)} = \frac{6.7^k}{\text{linear foot of wall}}$$

Now that we have the pressures and forces acting on the wall, we can now calculate the force to resist against sliding and the overturning moment.

Summing the horizontal forces which will cause the wall to slide is simply as follows:

$$H_{\text{sliding}} = H_a + H_L = 7.52^k + 1.75^k = 9.27^k$$

and summing the moments which will cause the footing to overturn about the toe, is calculated as follows:

$$M_{overturning} = H_L(10.75') + H_a(7.17') = 1.75^k(10.75') + 7.52^k(7.17') = 72.73^{k \cdot ft}$$

The component weights of the wall are calculated to determine the resulting frictional force. The weights are calculated here and are performed in the same manner as the forces, that is, based on per foot of linear wall. Their locations are shown in Figure 10.47.

$$W_L = 0.3^{k/ft^2}(9') = 2.7^k$$

$$W_{Sa} = 0.120^{k/ft^3}(20')(9') = 21.6^k$$

$$W_{Sp} = 0.120^{k/ft^3}(3')(1') = 0.36^k$$

$$W_w = 0.150^{k/ft^3}(23')(1.5') = 5.18^k$$

$$W_b = 0.150^{k/ft^3}(13.5')(1.5') = 3.04^k$$

The frictional resistance is then calculated as follows:

$$F = \mu(W_L + W_{Sa} + W_{Sp} + W_w + W_b) = 0.4(32.88^k) = 13.15^k$$

Total resistance, against sliding, is the sum of the frictional force and the passive pressure.

$$H_{resistance} = F + H_p = 13.15^k + 6.7^k = 19.85^k$$

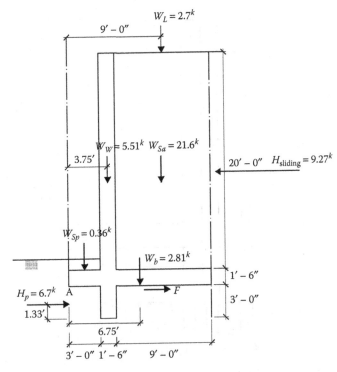

Figure 10.47 Illustration of horizontal and vertical loads on the retaining wall.

Then, the factor of safety for sliding is determined as follows:

$$\text{Factor of safety, F.S.} = \frac{H_{\text{resistance}}}{H_{\text{sliding}}} = \frac{19.85^k}{9.27^k} 2.14 > 1.5, \text{ therefore OK}$$

Next, calculate the resistance to overturning or resisting moment. The weight of the soil on top of the toe and the passive pressure is conservatively omitted in determining the overturning moment.

From Figure 10.47

$$+ \jmath \sum M_A = (W_L + W_{Sa})(9') + W_w(3.75') + W_b(6.75')$$

$$= (24.3^k)(9') + (5.51^k)(3.45') + (2.81^k)(6.75')$$

$$= 258.3^{k \cdot ft}$$

Then, the factor of safety against overturning is determined as follows:

$$\text{Factor of safety, F.S.} = \frac{M_R}{M_o} = \frac{258^{k \cdot ft}}{72.73^{k \cdot ft}} = 3.55 > 2.0, \text{ therefore OK}$$

Finally, to calculate the unfactored soil pressure at the base, as shown in Figure 10.48, we must determine the eccentricity on the footing/base.

Using the familiar eccentric footing procedure first, determine the kern of the footing.

$$e = \frac{L}{6} = \frac{13.5}{6} = 2.25'$$

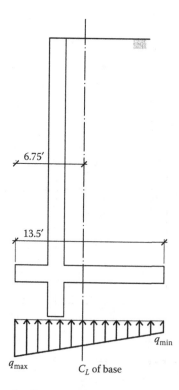

Figure 10.48 Bearing capacity stress distribution on bottom of footing.

Set an equilibrium equation to determine the center of gravity.

$$R\bar{x} = M_R - M_0$$

The resultant is the sum of the wall components.

$$R = W_L + W_w + W_{Sa} + W_b = 2.7^k + 5.51^k + 21.6^k + 2.81^k = 32.62^k$$

Solve for $\bar{x}$:

$$\bar{x} = \frac{M_R - M_0}{R} = \frac{258^{k\text{-ft}} - 72.73^{k\text{-ft}}}{32.62^k} = 5.68'$$

then,

$$e' = \bar{x}$$

and

$$e = 6.75' - 5.68' = 1.07' < \frac{L}{6}$$

Thus, the footing/base has full bearing along the bottom of the footing and use Equation 10.6 to calculate the maximum and minimum unfactored bearing pressures.

$$q_{max/min} = \frac{R\left(1 \pm \dfrac{6e}{L}\right)}{(L)(B)} = \frac{32.62^k\left(1 \pm \dfrac{6(1.07')}{13.5'}\right)}{(13.5')(1')}$$

$q_{max} = 3.56^k/\text{linear foot of wall}$
$q_{min} = 1.27^k/\text{linear foot of wall}$

Chapter 11

Structural review of construction

11.1 CONSTRUCTION ADMINISTRATION

Construction administration is the process by which the design professionals, contractors, owner, and building official work together to insure the project is constructed in compliance with the construction contract documents and that materials and quality of workmanship meets acceptable industry standards. Each of the members of the team has their own specific role and responsibility, which they bring to the process. Their efforts are coordinated by the lead design professional, usually the architect of the project. Periodic meetings are set to review the progress of construction performed and to coordinate future efforts. The general contractor or the construction manager coordinates and facilitates the onsite meetings.

The structural engineer's key responsibility is the interpretation of structural drawings and specifications, the establishment of standards of acceptable workmanship in accordance with the adopted building codes, and the site observations of construction, for the purpose of determining whether the work is in general accordance with the construction contract documents.

Services of the structural engineer may extend to include in part or all of the following:

- Responding to requests for information that are consistent with the requirements of the structural drawings and specification (contract documents).
- Interpreting the contract documents.
- Preparing supplemental drawings as appropriate.
- Making site visits at intervals appropriate to the stage of construction.
- Reporting on known deviations from the construction documents and observed defects and deficiencies in the work.
- Rejecting work that does not conform to the construction documents.
- Reviewing requests for payments submitted by the contractor.

11.2 INSPECTIONS AND OBSERVATIONS

Beyond the site inspection to review progress of construction where the engineer may review general compliance and general progress of construction and beyond inspections performed by the building official, the International Building Code (IBC), requires specific types of inspections be made and properly documented by an approved agency certified to supply such inspections. The specific types of inspections, tests, and protocols as well as the requirements for structural inspections and observations are described in Chapter 17, Special Inspections and Tests, of the IBC.

11.2.1 Special inspector agency

Special inspections and tests are performed during construction by an approved special inspection agency to be provided by the owner. The approved agency must be made known to the building official having jurisdiction prior to commencement of performing inspections. In addition, the contractor cannot provide the agency, as it is considered to be a conflict of interest.

11.2.2 Certification of special inspection agency

Prior to the start of construction, the approved special inspection agency must provide written documentation (certification) to the building official demonstrating the competence and relevant experience and training of the special inspectors who will be performing the special inspections.

11.2.3 Eligibility to perform special inspections

The registered design professional in responsible charge and engineers of record involved in the design of the project are permitted to act as the approved agency and their personnel are permitted to act as special inspectors for the work designed by them, provided they qualify as special inspectors.

11.2.4 Documentation of inspections

The approved special inspection agency must keep records of special inspections and tests performed. The special inspector must submit reports of the special inspections and tests to the building official and to the registered design professional in responsible charge. Reports shall indicate the work inspected or tested was or was not completed in conformance to approved construction documents. Discrepancies shall be brought to the immediate attention of the contractor for correction. If they are not corrected the discrepancies shall be brought to the attention of the building official and to the registered design professional in responsible charge prior to the completion of that phase of the work. The approved agency is to submit a final report, documenting required special inspection and tests with corrections of discrepancies noted, to the building official for record.

11.2.5 Special inspection statement

Where special inspection and or testing are required, the registered design professional in responsible charge shall prepare a statement of special inspections. The statement of special inspections shall identify the following:

1. The materials, systems, components, and work required to have special inspection or testing by the building official or by the registered design professional in responsible for each portion of the work
2. The type and extent of each special inspection
3. The type and extent of each test
4. Additional requirements for special inspection or testing for seismic or wind resistance
5. For each type of special inspection, identification as to whether it will be continuous special inspection or periodic special inspection

11.2.6 Contractor's responsibility

The contractor responsible for constructing the main wind-force or seismic-force-resisting system is required to prepare a written statement of responsibility, which must contain the acknowledgement of awareness of the special requirements specified in the statement of special inspections. This statement must be submitted to the owner and to the building official prior to the commencement of work.

11.2.7 Structural observations

For structures sited in high seismic and wind regions, structural observations can be required to be performed to view the construction of seismic and wind structural systems. This requirement of structural observations is made in addition to the special inspections required for the project. The structural observer must be a design professional employed and provided by the owner to conduct structural observations at a frequency and extent set by the observer. The observer must submit a statement identifying the frequency and duration of the observations to the building official prior to the commencement of the site visits.

Structural observations for seismic resistance
   Structural observations shall be provided for those structures in seismic design category D, E or F, where one or more of the following conditions exist:
   a. The structure is classified as risk category III or IV
   b. The height of the structure is greater than 75 ft
   c. The structure is assigned to seismic category E, is classified as risk I or II, and is greater than two stories above grade
   d. When designated by the registered design professional in responsible charge of the structural design
   e. When required by the building official
Structural observations for wind requirements
   Structural observations shall be provided for those structures sited where the nominal design wind speed as determined in Section 1609.3.1, IBC, exceeds 110 mph, where one or more of the following conditions exist:
   a. The structure is classified as risk category III or IV
   b. The building height is greater than 75 ft
   c. When designated by the registered design professional in responsible charge of the structural design
   d. When required by the building official

At the conclusion of the project or work specified in the structural observation statement, a written report, prepared by the structural observer is required to be submitted to the building official, certifying that the observations have been completed and any deficiencies noted have been remediated to the best of the structural observer's knowledge.

11.2.8 Required special inspections and tests

Special inspections and tests are required for steel, concrete, masonry and wood construction, soils, deep foundations, fabricated components and fire resistant coatings as well as structures meeting required wind and seismic resistance. The categories of required special inspections and testing listed in Section 1705, of the IBC, and their requirements is summarized and provided here. For a complete and thorough review of required special inspections, the reader should see Chapter 17 of the IBC.

Steel construction

Special inspections and nondestructive testing of structural steel elements in buildings, structures and portions thereof shall be in accordance with the quality assurance inspection requirements of AISC 360

Cold-formed steel deck

In accordance with the quality assurance inspection requirements SDI QA/QC

Open-web steel joists and joist girders

In accordance with Table 11.1 (Table 1705.2.3, IBC)

Cold-formed steel trusses spanning 60 ft or greater

The special inspector is required to verify that the temporary and permanent bracing is installed as per approved truss submittal

Concrete construction

Special inspections and tests shall be performed in accordance with Table 11.2 (Table 1705.3)

Welding of reinforcing bars

In accordance with the requirements of AWS D.1.4

Material tests

In accordance with quality standards for materials in Chapters 19 and 20 of ACI 318

Masonry construction

Special inspections and tests of masonry construction shall be performed in accordance with the quality assurance program requirements of TMS 402/ACI530/ASCE 5 and TMS602/ACI 530.1/ASCE 6

Wood construction

Prefabricated structural wood elements, which are load bearing or are lateral load-resisting members or assemblies require special inspections. Special inspections of prefabricated structural wood components are to be conducted at the fabricator's shop during fabrication. However, a fabricator can be registered and approved by an approved special inspection agency to perform work on their premises. For a fabricator to maintain their certification, an approved agency must provide periodic auditing of the fabricator's procedural and quality controls manual as well as facility.

Structural wood elements, which are built on-site:

High-load diaphragms

The special inspector confirm the sheathing thickness and grade, framing members spacing and dimensions, the nail or staple size and number of fasteners of the built assembly is as conveyed by the approved construction documents.

Table 11.1 Required special inspections of open-web steel joists and joist girders

Type	Continuous special inspection	Periodic special inspection	Referenced standard
1. Installation of open-web steel joists and joist girders.			
a. End connections—welding or bolted.	–	X	SJI specifications
b. Bridging—horizontal or diagonal.	–		
1. Standard bridging.	–	X	SJI specifications
2. Bridging that differs from the SJI specifications.		X	

Table 11.2 Required special inspections and tests of concrete construction

Type	Continuous special inspection	Periodic special inspection	Referenced standard	IBC reference
1. Inspect reinforcement, including prestressing tendons, and verify placement.	–	X	ACI318: Ch. 20, 25.2, 25.3, 26.6.1–26.6.3	1908.4
2. Reinforcing bar welding:			AWSD1.4 ACI318; 26.6.4	–
a. Verify weldability of reinforcing bars other than ASTM A706;	–	X		
b. Inspect single-pass fillet welds, maximum 5/16"; and		X		
c. Inspect all other welds.	X			
3. Inspect anchors cast in concrete.	–	X	ACI318; 17.8.2	–
4. Inspect anchors postinstalled in hardened concrete members.			ACI318; 17.8.2.4 ACI318; 17.8.2	–
a. Adhesive anchors installed in horizontally or upwardly inclined orientations to resist sustained tension loads.	X			
b. Mechanical anchors and adhesive anchors not defined in 4.a.		X		
5. Verify use of required design mix.	–	X	ACI318: Ch. 19, 16.4.3, 26.4.4	1904.1, 1904.2, 1908.2, 1908.3
6. Prior to concrete placement, fabricate specimens for strength test, perform slump, and air content tests and determine the temperature of the concrete.	X	–	ASTM C172 ASTM C31 ACI318: 26.4, 26.12	1908.10
7. Inspect concrete and shotcrete placement for proper application techniques.	X	–	ACI318: 26.5	1908.6, 1908.6, 1908.8
8. Verify maintenance of specified curing temperature and techniques.	–	X	ACI318: 26.5.3–26.5.5	1908.9
9. Inspect prestressed concrete for:			ACI 318: 26.10	–
a. Application of prestressing forces; and	X	–		
b. Grouting of bonded prestressing tendons.	X	–		
10. Inspect erection of precast concrete members.	–	X	ACI318: Ch. 26.8	–
11. Verify in situ concrete strength, prior to stressing of tendons in post-tensioned concrete and prior to removal of shores and forms from beams and structural slab.	–	X	ACI318: 26.11.2	–
12. Inspect formwork for shape, location, and dimensions of the concrete member being formed.	–	X	ACI318: 26.11.2(b)	–

Metal-plate-connected wood truss spanning 60 ft or greater
The special inspector is required to verify that the temporary and permanent bracing is installed as per approved truss submittal

Soils
In accordance with Table 11.3 (Table 1705.6, IBC)

Driven deep foundations
In accordance with Table 11.4 (Table 1705.7, IBC)

Table 11.3 Required special inspections and tests of soils

Type	Continuous special inspection	Periodic special inspection
1. Verify materials below shallow foundations are adequate to achieve the design bearing capacity.	–	X
2. Verify excavations are extended to proper depth and have reached proper material.	–	X
3. Perform classification and testing of compacted fill materials.	–	X
4. Verify use of proper materials, densities, and lift thicknesses during placement and compaction of compacted fill.	X	–
5. Prior to placement of compacted fill, inspect subgrade and verify that site has been prepared properly.	–	X

Table 11.4 Required special inspections and tests of driven deep foundation elements

Type	Continuous special inspection	Periodic special inspection
1. Verify element materials, sizes, and lengths comply with the requirements.	X	–
2. Determine capacities of test elements and conduct additional load tests, as required.	X	–
3. Inspect driving operations and maintain complete and accurate records for each element.	X	–
4. Verify placement locations and plumbness, confirm type and size of hammer, record number of blows per foot of penetration, determine required penetrations to achieve design capacity, record tip and butt elevations, and document any damage to foundation element.	X	–
5. For steel elements, perform additional special inspections in accordance with Section 1705.2.	–	–
6. For concrete elements and concrete-filled elements, perform tests and additional special inspections in accordance with Section 1705.3.	–	–
7. For specialty elements, perform additional inspections as determined by the registered design professional in responsible charge.	–	–

Cast-in-place deep foundations
In accordance with Table 11.5 (Table 1705.8, IBC)
Helical pile foundations
Continuous special inspections shall be performed during installation of helical piles
foundations

Table II.5 Required special inspections and tests of cast-in-place deep foundation elements

Type	Continuous special inspection	Periodic special inspection
1. Inspect drilling operations and maintain complete and accurate records for each element.	X	–
2. Verify placement locations and plumbness, confirm element diameters, bell diameters (if applicable), lengths, embedment into bedrock (if applicable), and adequate end-bearing strata capacity. Record concrete or grout volumes.	X	–

Fabricated items

Special inspections of prefabricated structural components are to be conducted at the fabricator's shop during fabrication. However, a fabricator can be registered and approved by an approved special inspection agency to perform work on their premises. For a fabricator to maintain their certification, an approved agency must provide periodic auditing of the fabricator's procedural and quality controls manual as well as facility.

Special inspections for wind resistance

Special inspections for wind resistance are required for buildings and structures constructed in the following areas:

a. In wind exposure category B, where the nominal design wind speed as determined in Section 1609.3.1, IBC, is 120 mph or greater

b. In wind exposure category C or D, where the nominal design wind speed as determined in Section 1609.3.1, IBC, is 110 mph or greater

Structural wood

Continuous special inspection is required during field gluing operations of elements of the main wind force-resisting system. Periodic special inspection is required for nailing, bolting anchoring, and other fastening of elements of the main wind force-resisting system.

Cold-formed steel light-frame construction

Periodic special inspection is required for welding operations of elements of the main wind force-resisting system

Periodic special inspection is required for screw attachment, bolting anchoring, and other fastening of elements of the main wind force-resisting system

Wind-resisting components

Periodic special inspection is required for fastening of the following systems and components:

i. Roof covering, roof deck, and roof framing connections

ii. Exterior wall covering and wall connections to roof and floor diaphragms and framing

Special inspections for seismic resistance

Special inspections for seismic resistance are required as follows:

Structural steel

Seismic-force-resisting system

Buildings and structures assigned to seismic design category B, C, D, E or F shall be performed in accordance with the quality assurance requirements of AISC 341.

Structural wood

For the seismic-force-resisting systems of structures assigned to seismic design category C, D, E or F, continuous special inspection is required during field gluing operations of elements of the main wind force-resisting system. Periodic special inspection is required for nailing, bolting anchoring, and other fastening of elements of the main wind force-resisting system.

Cold-formed steel light-frame construction

For the seismic-force-resisting systems of structures assigned to seismic design category C, D, E, or F, periodic special inspection is required for welding operations of elements of the main wind force-resisting system.

Periodic special inspection is required for screw attachment, bolting anchoring, and other fastening of elements of the main wind force-resisting system.

Codes and Bibliography

CODES

Building Code Requirements and Specifications for Masonry Structures (and companion commentaries), 2013, The Masonry Society, Boulder, CO; American Concrete Institute, Detroit, MI; and Structural Engineering Institute of the American Society of Civil Engineers, Reston, VA. [TMS 402/602 2013]

Building Code Requirements for Structural Concrete, 2011, American Concrete Institute, Farmington Hills, MI. [ACI 318-11]

International Building Code, 2015 edition, International Code Council, Falls Church, VA. [IBC 2015]

Minimum Design Loads for Buildings and Other Structures, 2010, American Society of Civil Engineers, Reston, VA. [ASCE 7-10]

National Design Specification for Wood Construction ASD/LRFD, 2012 edition & *National Design Specification Supplement, Design Values for Wood Construction*, 2012 edition, American Forest & Paper Association, Washington, DC. [NDS 2012]

SEAOC Structural/Seismic Design Manual (2009 IBC).

Seismic Design Manual, 2nd edition, American Institute of Steel Construction, Chicago, IL. [AISC SDM 2014]

Special Design Provisions for Wind and Seismic with Commentary, 2008 edition, American Forest & Paper Association, Washington, DC. [NDS SDPWS 2008]

Steel Construction Manual, 14th edition, American Institute of Steel Construction, Chicago, IL. [AISC 2014]

BIBLIOGRAPHY

American Institute of Architects, *The Architect's Handbook of Professional Practice*, 14th Ed., John Wiley & Sons, Inc., Hoboken, NJ, 2008.

Amrhein, J.E., *Reinforced Masonry Engineering Handbook*, 5th Ed., Masonry Institute of America, Los Angeles, CA, 1998.

Concrete Reinforcing Steel Institute (CRSI), *Design Handbook 2008*, 10th Ed., Concrete Reinforcing Steel Institute, Schaumburg, IL, 2008.

Levinson, I.J., *Statics and Strength of Material*, Prentice-Hall, Upper Saddle River, NJ, 1971.

Lindeburg, M.R., McMullin, K.M., *Seismic Design of Building Structures*, 9th Ed., Professional Publications, Belmont, CA, 2008.

Liu, C., Evett, J.B., *Soils and Foundations*, 2nd Ed., Prentice-Hall, Upper Saddle River, NJ, 1987.

O'Rourke, M., *Snow Loads: Guide to Snow Load Provisions of ASCE 7-05*, ASCE Press, Reston, VA, 2007.

Riley, W.F., Sturges, L.D., Morris, D.H., *Statics and Mechanics of Materials*, John Willey & Sons, New York, 1995.

Spillers, W.R., *Introduction to Structures*, 2nd Ed., Horwood Publishing Limited, West Sussex, England, 2002.

Williams, A., *Structural Engineering Reference Manual*, 6th Ed., Professional Publications, Belmont, CA, 2012.

Index

Note: Page numbers followed by f and t refer to figures and tables, respectively.

Printed in the United States
by Baker & Taylor Publisher Services